EARTH SCIENCES IN THE 21ST CENTURY

BASINS

METHODS OF FORMATION, ONGOING DEVELOPMENTS AND EMERGING CHALLENGES

EARTH SCIENCES IN THE 21ST CENTURY

Additional books in this series can be found on Nova's website
under the Series tab.

Additional e-books in this series can be found on Nova's website
under the e-book tab.

EARTH SCIENCES IN THE 21ST CENTURY

BASINS

METHODS OF FORMATION, ONGOING DEVELOPMENTS AND EMERGING CHALLENGES

JIANWEN YANG
EDITOR

New York

For permission to use material from this book please contact us:
Telephone 631-231-7269; Fax 631-231-8175
Web Site: http://www.novapublishers.com

Library of Congress Cataloging-in-Publication Data

ISBN: 978-1-63117-510-7

Published by Nova Science Publishers, Inc. † New York

Contents

PREFACE

Sedimentary basins represent the vast repositories for the Earth's natural resources, containing not only a large variety of metallic mineral deposits, but also natural gas, oil, oil sands and coal deposits; and therefore are a continuous focus of research. The formation of these economic resources in sedimentary basins results from complex interactions among geological, geochemical, thermal, structural and hydrological processes. This book is composed of 7 chapters written by 21 authors from 6 countries (Argentina, Brazil, Canada, China, Ghana, and Nigeria), reflecting the diversity of topical case studies in various regions. The collection of topics aims to present recent advances in research on basins and the related issues, including Tectonic evolution of Youjiang basin and the resultant Carlin-type gold deposits (Chapter 1); Numerical simulation of turbidity current and reservoir prediction in Qiongdongnan basin of northern South China Sea (Chapter 2); Similarities and differences between 2-D and 3-D numerical results of ore-forming fluid flow in sedimentary basins (Chapter 3); Subsidence, hydrocarbon generation and thermal history modelling of Chad basin (Chapter 4); Hydrological case study of flatlands in Argentina (Chapter 5); Hydrochemical study of surface waters in the Barreiro hydrographic basin, Brazil (Chapter 6); and Climate downscaling over Densu basin and nexus on hydrology in Ghana (Chapter 7).

Chapter 1 - Sedimentary rock-hosted Carlin-type gold deposits have been considered geologically distinct and economically significant, with a great exploration potential in China. The Youjiang Basin of southwestern China, as the famous Dian-Qian-Gui 'Gold Triangle', hosts many large gold deposits of this type. However their origin remains unclear and highly controversial. Based on the recent geological and geochemical data, the authors speculate that the Carlin-type gold deposits in this region may not be the product of a single-sourced fluid, but probably due to the multi-sourced fluids occurring at many periods and stages corresponding to the basin evolution. The authors therefore establish an ore-forming model, based on the sedimentary tectonic evolution, that contains sedimentary diagenetic stage, orogenic metamorphic stage, and postorogenic extension and magmatic activity stage.

Chapter 2 - The hydrocarbon explorations in deepwater areas around the world have achieved great success in past decades. The main targets of deepwater exploration are submarine fans. But some of them comprise shale or mud and therefore cannot be hydrocarbon reservoirs. Therefore, the reservoir prediction before drilling is essential and important for hydrocarbon exploration in deepwater areas. Unfortunately, the wells are few in deepwater areas, so they cannot be used to make effective predictions by conventional log-constrained inversion. Hydrodynamic simulation can illuminate the sedimentary processes of gravity flows, including turbidite currents, which are the main geneses of submarine fans. Combined with the geological background in the Qiongdongnan basin (QDNB), northern South China Sea, this study simulates the sedimentary processes and geometric shapes of turbidites composed of various size particles. Then, compared with seismic reflection features, the main component of the submarine fans interpreted in seismic data is estimated. The prediction has been confirmed by several recent drillings in the QDNB. Moreover, the results show that the slope gradient controls the development of turbidity currents and 1.5°-3° are more appropriate for triggering turbidity current that can flow a long distance along the slope. The turbidities are usually deposited as slope fans and basin floor fans, and the single turbidite has the thickest segment in the slope foot and thins toward the basin. The authors conducted the simulation with various grain sizes and observed the geometric shapes of formed submarine fans, and found that the coarse grain sizes generate thick turbidities with a limited spread, thicker and smaller, and vice versa.

Chapter 3 - This chapter aims at addressing the major similarities and differences between 2-D and 3-D numerical modeling results of ore-forming hydrothermal fluid flow in sedimentary basins, with Mount Isa Basin, northern Australia as an example. Our numerical results of Sedex-type deposits indicate that both the 2-D and 3-D fluid flow and hydrothermal discharge are controlled by the spatial relationships between active synsedimentary faults and clastic aquifers. However, the general recharge-discharge pattern of fluid flow established on the basis of 2-D modeling is usually not valid under 3-D conditions, even for a quasi 3-D domain constructed simply by extrapolating the 2-D cross-section along the longitudinal direction. This is because fluid tends to circulate within more permeable fault zones and form a series of planar convection cells over the fault planes rather than in less permeable host rocks, unless the faults and host rocks have a similar permeability range. Including a secondary cross fault to the 3-D model makes the fluid flow pattern as well as the heat regime become more distinct from the 2-D modeling result. Localized, subvertical columns of enhanced permeability, related to the intersection of the primary major faults with the secondary cross fault, are essential to developing 3-D 'mushroom-shaped' hydrothermal convection rolls that allow significant amount of fluids to circulate through the host rocks and leach sufficient metal content to form deposits of this type. Salinity variation through a basin has important implications for hydrothermal fluid flow, either promoting or impeding buoyancy-driven free convection. Our modeling results indicate that Sedex-type deposits are more easily formed when evaporation first produces surface brines and then these brines sink and displace pore waters in basins.

Chapter 4 - Subsidence, burial and thermal history studies are crucial factors in determining the hydrocarbon potential (both quality and quantity) of a basin. In this study subsidence and thermal analyses of Chad Basin Nigeria were investigated utilizing logs data from 23 exploratory wells. The oceanic lithospheric cooling concept was used for the basin subsidence while the sedimentation history in the basin was reconstructed utilizing the method of "back-stripping" at different locations. The results of the thermal and subsidence analyses were used to model the hydrocarbon maturation levels in different stratigraphic units of interest. Subsidence and sedimentation were observed to have occurred mostly in the Albian to Maastrichtian times, and only the sediments from late Santonian units in the basin have reached sufficient thermal maturity (at a depth range of 1.5 to 3.8 km) to generate hydrocarbons. The hydrocarbon may still be occurring at present day. Maturation in terms of virtrinite reflectance (Ro) and maturation index (C) ranges from 0.6 and 1.37 percent, and 9.9 to 15.2, respectively, for these matured sediments. Time-temperature relation has been demonstrated to be a controlling factor for diagenetic and organic metamorphic processes in rocks. Level of maturation of organic matter was expressed using various maturation indices such as paleotemperature, oil window concept and virtrinite reflectance. The thermal model results show that basement subsidence is a function of the square root of the age of the basin, indicating that subsidence in Chad Basin Nigeria can be explained by a simple thermal contraction of the lithosphere following an extensional phase. With the results of the various modelling indices, the study has revealed the prospect of the basin to generate hydrocarbons in a commercial quantity.

Chapter 5 - A good deal of the Earth's population inhabits large flatlands that very often supply huge quantities of food to humankind. Topographic slope of the hydrologic system of a plain is very small, or zero, which is sometimes insufficient to generate an integrated drainage network, a surface runoff or to cause fluvial erosive processes. A sector of the Pampas Plain, locally known as "Pampa Arenosa" (Sandy Pampa), in the northwest of the Buenos Aires Province, Argentina, is presented as a case study. With an area of about 50,000 km2, this region presents a monotonous plain surface of low relief. It is a landscape shaped by eolic processes where it is possible to distinguish different paleoforms related to dune ridges. Current wet climate conditions contrast with the origin of an eolic landscape associated with a preceding arid climate. The dominant features are the absence of a drainage network and the accumulation of water in blowouts or in inter dune openings. The general characteristics of the region, as well as the hydrologic factors (rainfall, water surplus, soil wetness, water table level) are analyzed, which reveals the most marked impacts from alternate flood and dry periods. The way in which the regional hydrologic variations affect the agricultural production system in one of the most important agricultural regions of Argentina is also studied. The main feature of a large flatland is its low morphogenetic energy, which makes the energy of water resources dissipate through vertical water movements (evaporation, transpiration, infiltration, interchanges between the unsaturated zone and the water table). These vertical movements predominate over surface or groundwater horizontal movements, thus enhancing the importance of surface and groundwater storage variations. It follows that flatlands are highly sensitive to climatic fluctuations (water surplus and deficit), as well as to human activities. These characteristics make it difficult to effectively apply the traditional concept of a basin used for regions having an abrupt relief. Moreover, the difficulty to define a study area as a drainage basin leads to a need for replacing the term drainage basin by hydrologic region. In addition, the development of methodologies for quantification, measurement and hydrologic forecasting adapted to this kind of regions is also a challenging task.

Chapter 6 - The chemical evolution of the Earth´s surface and, consequently, of the hydrographic basins, is controlled by several factors, among which are the interaction of rainwaters, the atmosphere and the continental crust. Different climatic conditions affect the denudation, chemical weathering and physical erosion in hydrographic basins. The present and future design of the hydrographic basins is affected by chemical weathering processes and such information has been increasingly used by decision makers engaged in the management of our river basin systems. A lot of factors have been applied in models that utilize the concentrations of constituents like sodium, calcium, potassium, magnesium and total dissolved solids for estimating the chemical weathering fluxes taking place in the drainage basins. On the other hand, the population growth in all countries has caused an increased concern on the quantity and quality of the water resources in hydrographic basins. Well-defined anthropogenic inputs and/or pollution from diffuse source loads have commonly affected the chemistry of water bodies, requiring special attention by people involved in the river basin management. This chapter reports the results of a hydrochemical study held at the Barreiro hydrographic basin, Araxá city, Minas Gerais State, Brazil. The surface water quality in the site is not only affected by geogenic inputs but also by anthropogenic sources that are mainly coupled to activities involving the mining and production of fertilizers, which take place in alkaline rocks belonging to the geological formations of the region.

Chapter 7 - Climate Downscaling is a term adopted in climate science in recent years to describe a set of techniques that relate local to regional-scale climate variables in relation to the larger scale atmospheric forcing. Theoretically, the techniques are advancements of the known traditional techniques in synoptic climatology. Climate downscaling specifically addresses the detailed temporal and spatial information from Global Climate Models (GCMs) required by precise researches of today. The Regional Climate Model version 4 (RegCM4), with a horizontal resolution of 55 km, was used to downscale the ECHAM5 simulations forced with observed SSTs over southern Ghana. An ensemble of 10 ECHAM5 AGCM integrations forced with observed time-evolving SSTs was done from 1961 to 2000. For each of the ECHAM5 AGCM integrations, a nested integration with the REGCM was done for the period January–June 1961–2000. Six-hour wind, temperature, humidity, and surface pressure data from ECHAM45 AGCM outputs were linearly interpolated in time and in space onto REGCM grids as base fields. The results of the comparison for the Densu catchment station showed a good correlation between the observed REGCM-simulated monthly rainfalls with significant statistics. Although no coherent trends were found in the basin, interannual rainfall variability was more pronounced as revealed by the REGCM 4 simulations. The northern part of the basin is most vulnerable to these variations because it has a monomodal rainfall pattern compared to the south which has relatively higher rainfall amounts due to its bi-modal rainfall pattern. The SPI analysis conducted on projected precipitation based on REGCM using IPCC's A1B and B1 scenarios against the base period of 1961-2000 showed both scenarios agreeing to a general drying trend for the future.

About the Editor

Jianwen Yang[1,2]
[1] College of Earth Sciences
Guilin University of Technology
Guilin, Guangxi, China, 541004
[2] Department of Earth and Environmental Sciences
University of Windsor
Windsor, Ontario, N9B 3P4, Canada
Email: jianweny@uwindsor.ca
Tel: (519)253-3000/ext 2181
Fax: (519)973-7081

Jianwen Yang received a Master's degree and a Ph.D. degree from the University of Toronto, Canada, and has worked in China, Australia and Canada. Dr. Yang is an Adjunct Professor at the College of Earth Sciences, Guilin University of Technology, and a full-time Professor at the Department of Earth and Environmental Sciences, University of Windsor, specializing in lab- and field-based computational numerical modeling of subsurface fluid flow, heat and mass transport, and chemical reactions associated with applications in earth and environmental sciences.

In: Basins
Editor: Jianwen Yang
ISBN: 978-1-63117-510-7

Chapter 1

SEDIMENTARY TECTONIC EVOLUTION OF YOUJIANG BASIN AND FORMATION OF CARLIN-TYPE GOLD DEPOSITS, SOUTHWESTERN CHINA

Baocheng Pang[1,*], Xijun Liu[1], Jianwen Yang[1,2], Hai Xiao[1] and Wei Fu[1]

[1]College of Earth Sciences, Guilin University of Technology, Guilin, Guangxi, P.R. China
[2]Department of Earth and Environmental Sciences, University of Windsor, Windsor, Ontario, Canada

ABSTRACT

Sedimentary rock-hosted Carlin-type gold deposits have been considered geologically distinct and economically significant, with a great exploration potential in China. The Youjiang Basin of southwestern China, as the famous Dian-Qian-Gui 'Gold Triangle', hosts many large gold deposits of this type. However their origin remains unclear and highly controversial. Based on the recent geological and geochemical data, we speculate that the Carlin-type gold deposits in this region may not be the product of a single-sourced fluid, but probably due to the multi-sourced fluids occurring at many periods and stages corresponding

* Corresponding author: Email: pbc@glut.edu.cn.

to the basin evolution. We therefore establish an ore-forming model, based on the sedimentary tectonic evolution, that contains sedimentary diagenetic stage, orogenic metamorphic stage, and postorogenic extension and magmatic activity stage.

Keywords: Carlin-type gold deposit, Youjiang basin, basin evolution, ore-forming model, hydrothermal fluid flow

1. INTRODUCTION

Carlin-type gold deposit is a new type of gold after its first discovery in Nevada of the United States in 1961 (Cline et al., 2005). Since then it has been found in China, Australia, Dominican Republic, Spain, Russia, Malaysia, Philippines, Yugoslavia, Greece, Iran, and Indonesia (Li. et al., 1998; Cline et al., 2005). The northern Nevada and the southwestern China represent the two largest Carlin-type gold regions in the world. The gold reserves discovered in the United States are more than 6,000 tons, making the USA the second largest gold concentrated region of the world (Muntean et al., 2011), just behind the South Africa. The gold reserves that have been found on the margin of Yangtze Block of southwestern China, including the two Golden Triangle regions of West Qinling belt and Youjiang basin, are more than 2,000 tons, representing an important gold resource base of China. Though the proven reserves are less than those of the United States, southwestern China is very prosperous in finding more deposits of this type.

The Youjiang basin on the southwestern margin of Yangtze block of South China is located at the junction of southwestern Guizhou, northwestern Guangxi and southeastern Yunnan (Figure 1). Due to the formation of many large Carlin-type gold deposits and other polymetallic deposits, and also as the potential area of oil and gas resources in South China, it has drawn great attention of Chinese and overseas geologists. Over the past three decades, Chinese and foreign scholars have conducted extensive and intensive studies on the Carlin-type gold deposits in this region. However, the origin of ore is still highly controversial. The ore-forming models proposed previously include meteoric water (Li et al., 1989; Guo et al., 1992; Pang et al., 2001; Hu et al.., 2002), magmatic hydrothermal (Liu et al., 1996; Zhu et al., 1998; Wang, 1999, 2000; Su et al., 2001), Basinal fluids (Liu et al., 1997; Liu et al., 1997; Liu et al., 2002; Wang et al., 2002; Gu et al., 2007; Gu et al., 2011), metamorphic fluids (Hofstra et al., 2005; Su et al., 2009a), Tectonometallogenic (Peters et al., 2007; Chen et al., 2011), and more. Also, a consensus is not reached on the timing of ore formation. The existing Chronology spans a wide time period from 267-62 Ma (Wang et al., 1989; Zhang et al., 1992; Hu et al., 1995; Su et al., 1998; Chen et al., 2007; Su et al., 2009). Although most of the data show that the ore likely formed in the Yanshan period, other periods are still possible.

There has been a large debate on the genesis of Carlin-type gold deposits. This may be due to the existence of different deposits in a same region and similar deposits in different regions in terms of geological and geochemical characteristics. Or perhaps this is due to the diversity of geological and geochemical characteristics produced by complicated ore-forming processes in a same deposit.

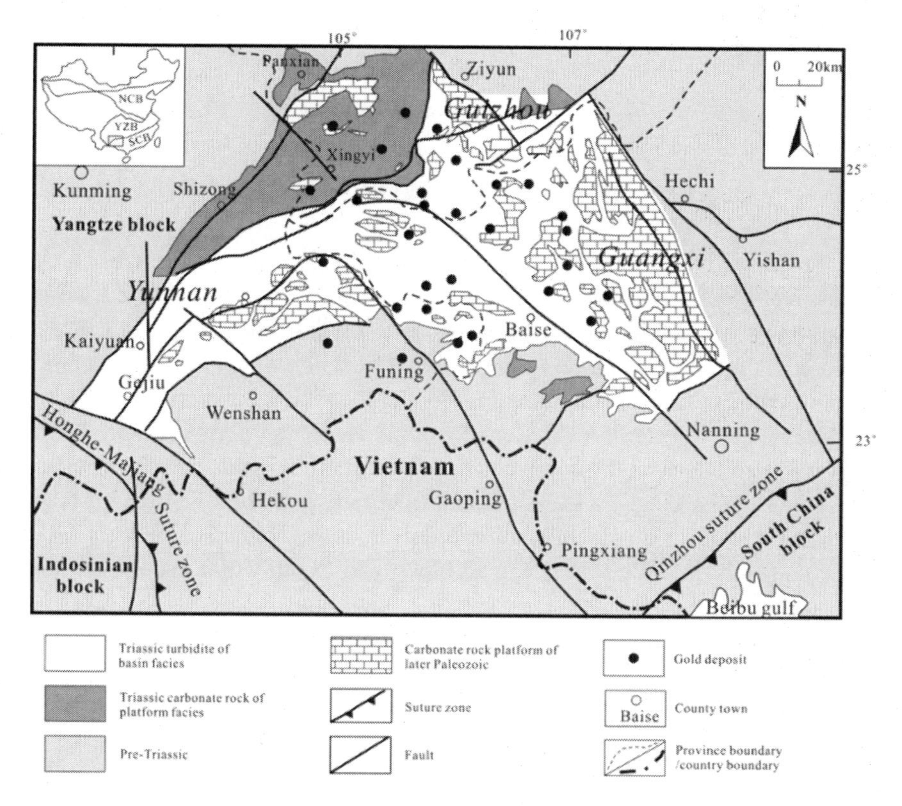

Figure 1. A simplified regional geological map showing the location of Carlin-type gold deposits in Yunnan-Guizhou-Guangxi "Golden Triangle" region. NCB=North China Block; YZB=Yangzi Block; SCB = South China Block (Modified after Chen et al., 2011 and Du et al., 2009).

This may also be related to the data inconsistencies caused by the differences in research time, research objectives and the methods employed. Gold deposits in different regions may have similar geological and geochemical characteristics, whereas gold deposits in adjacent regions sometimes exhibit different characteristics, indicating that these gold deposits may not form from a single-source fluid, but more likely from multi-source fluids involving a couple of periods and stages. They may be the product of different tectonic evolutions under an identical geological background. Therefore, it is necessary to analyze the reasons and processes for the formation of Carlin-type gold deposits from the point of view of sedimentary tectonic evolution of a basin, which may help explain some of inconsistent data. It should be noted that in Youjiang basin, in addition to the abundant Carlin-type gold deposits, there also exist manganese, tin, lead, zinc, antimony, barite, and siderite deposits. This article will only address the relationship between the basin evolution and gold deposits.

2. Sedimentary Tectonic Evolution

Youjiang basin is located in the transition zone of Yangtze block, Indosinian block and South China block. It is nearly a rhomb-shaped basin, and roughly bounded by Mile-Shizong fault, Ziyun-Yishan fault, Red River fault, and Qinzhou fault. Due to the double effects of Circum Pacific tectonic domain and Tethys tectonic domain, Youjiang basin has experienced a long and complicated tectonic evolution. During Caledonian period, South China took Yangtze Block as the continental nucleus, merging with Indosinian block and South China block in the south, and merging with Sino-Korean block in the north. However, this combination (equivalent to Guangxi movement) was not completed, in which the Yangtze block and South China block met and connected along the Jiangshan-Shaoxing fault zone, leaving the wedged Qinfang residual trough on its south (Shui, 1986). At the same time, the Indosinian block was also combined with these two blocks in the north along the Jinsha River-Red River (Ailaoshan Mountain) fault zone. As the arrival of Hercynian-Indosinian divergent-convergent tectonic cycles, a large-scale rifting activity occurred along the Yangtze block margin (Wang, 1985), forming a series of extensional belts along the continental margin (Zhu et al., 1992). Youjiang basin began its development and evolution on the southwestern margin of the Yangtze block under this geodynamic background.

Due to the superimposition of different tectonic domains, the evolution of Youjiang basin is very complicated. Geological community therefore has different views on the type and nature of the basin corresponding to different stages. For instance, Wu et al. (1990) proposed an evolution of foreland basin (D-P_1)-passive continental marginal rift basin (P_2-T_1)-foreland basin (T_2-T_3); whereas Zeng et al. (1995) proposed a different idea of rift basin of passive continental margin (D_1-P_1)-back-arc basin (P_2-T_3). In addition, there are also a variety of opinions such as early rift (D-P_1)-late rift and passive continental margin (P_2-T_1)-foreland basin (T_2-T_3) (Qin et al., 1996), as well as the rift basin of continental margin (D_1-C)-back-arc basin (P-T_1)-back-arc foreland basin (end of T_1 -T_2) (Du et al., 2009).

In spite of the inconsistence, most researchers agree that Youjiang basin experienced rifting on the edge of Yangtze continent during Devonian and closed at Late Triassic. According to the sedimentary filling pattern, the basin evolution can be divided into two stages. During the period of Early Devonian to early and middle stages of Early Triassic (D_1-_T_1^{a-b}), the sediments were mainly composed of thin dark limestone, siliceous rock, laminated mudstone and tuff, and often associated with debris flow sediments related to the carbonate construction of residual isolated platform. These are the characteristics of a rift basin on continental margin in the stretching stage. In the period of late stage of Early Triassic to Late Triassic (T_1^c - T_3), the basin was characterized by the filling sequences of Alpine foreland basin. That is, the upper part of Lower Triassic is made up of thin mudstone, and marl with siltstone and siliceous rock that formed under a deep water condition with little sediment inputs. The Middle Triassic is composed of very thick turbidites with flysch features. The Upper Triassic is made up of coarse clastic sediments and coal-bearing sediments with molasses feature. The strata thickness shows that the sedimentary center began to migrate since the Early Triassic to Middle Triassic towards the continental direction of north northeast, consistent with the direction of ancient flow during this period (Wang, 1985; Wu et al., 1993). Taking the basic pattern of Yangtze block and Tethys ocean into consideration, and combining the basement tectonic unit of Yangtze block and the kinetics of micro block in relation to the formation and evolution of Youjiang basin, the basin can be divided into the rift basin stage of passive continental margin (D_1-P_1), back-arc rift basin stage (P_2-T_1^{a-b}) and back-arc foreland basin stage (T_1^c -T_3) (Figure 2).

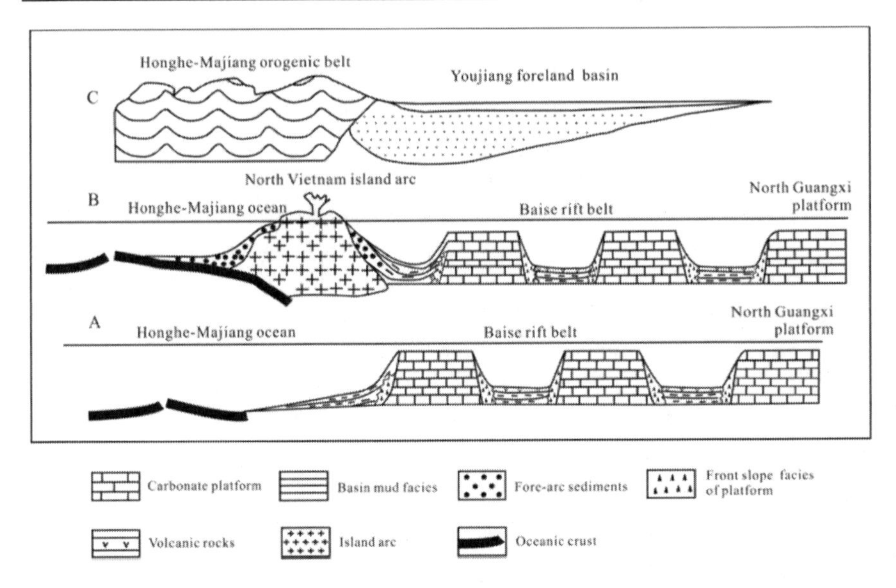

Figure 2. Tectonic evolution of Youjiang basin (Modified after Du et al., 2009). A. D_1-P_1 continental margin rift stage; B. P_2- T_1^{a-b} back-arc rift basin stage; C. T_1^{c}-T_3 back-arc foreland basin stage.

In general, Youjiang basin was open along the fault zone of Ailaoshan Honghe from the Early Devonian, leading to the appearance of an NW direction extensional basin parallel to the combining belt. During the whole period of Hercynian and Indosinian, Youjiang basin was an oceanic basin as a part of ancient Tethys ocean (Zeng et al., 1995). The Youjiang basin, in the context of appearing, developing and disappearing of ancient Tethys tectonic domain, has experienced an evolution from passive continental margin to foreland basin, and finally the orogeny collision formed Indosinian-Yanshan orogenic belt (Wang, 1994). The Indosinian movement was a strong orogenesis, which made Youjiang basin fold. The Yanshan movement caused very strong faulting and block disintegrating, and resulted in acidic magma intrusions, leading to the formation of small stocks, dikes and deeply buried igneous rocks (Chen, 2002).

3. Regional Geological Background

The shape and internal structure of Youjiang basin are controlled mainly by the NW direction major syngenetic faults and the NE direction secondary syngenetic faults, which constitutes the basin boundary. The internal faults occurred mostly in the period of Hercynian-Indosinian, and were restricted by the boundary faults. In the entire period of Hercynian-Indosinian, most of them were tensional and formed a number of grabens and horsts, which caused the basin not only to be diamond-shaped, but also to be divided by several secondary basins in its interior. Thus, the sedimentary characteristics of the basin is distinct from the paelogeographical framework of this region, characterized by an alternating distribution of deep water basins (depression areas) and shallow-water carbonate platforms (uplift areas).

Due to the supply difference of sediment sources caused by different tectonic environments, and also due to the change of depositional and settlement rate, the early deposition in the basin evolved from siliceous rocks of very small thickness to carbonates or calcium debris turbidites of larger thickness. In late period, it evolved again from siliceous rocks - volcano turbidites to epicontinental clastic turbidites of great thickness, which made the entire basin be overlapped by a double-layer structure composed of non-compensatory sediments and compensatory or super compensatory sediments with different characteristics. Among them, the lower layer is mainly composed of D-P carbonate; while the upper layer is mainly composed of T_1-T_2 clastic rocks (T_1 is the volcanic clastic turbidites and T_2 is the terrigenous clastic turbidites). These strata are widely distributed in the basin.

In particular, the Triassic turbidite is distributed nearly throughout the whole basin, and its thickness is generally more than 2000m with a maximum of over 6000m. The sedimentary centers are in Baise, Tianlin and Longlin. The turbidite stratum is composed of rhythmic greywacke and mudstone, and the thickness of rhythmic layer varies from 1cm to 5m. The rocks are mainly graywacke (mostly feldspar lithic graywacke and lithic quartz wacke), and secondarily are miscellaneous-powder sandstone, clay, shale and marl. Among them, the content of greywacke debris varies greatly, with a content of quartz of 20-75%, feldspar of 0-38%, debris of 1-50%, and matrix of >20%. The rock composition is complex and up to over ten types, including a variety of sedimentary rocks, volcanic rocks and granitic porphyry, and more, with a poor sorting, a low compositional maturity, and a low structural maturity. The particle size range of turbidite sandstones is within ϕ1-6, with an average of about ϕ3.2-3.7 (Guangxi Bureau of Geology and Mineral Resources, 1985). The turbidite generally has a clear Bauma sequence structure, which is mainly in the combination of sequence segments of AE, ABCDE, BCDE or CDE. A turbidite stratum with a complete Bauma sequence is typically 2-4m in thickness, with a maximum of up to 8m. The turbidite strata are stable with a large outspread, having a mutation at the bottom and a gradual change at the top. The turbidite particles tend to be thinner from the south to the north. The ratio of sand to mud is about 1:1. Notably, although the size of turbidite is fine, the porosity is big and the permeability is high in this area (Zhang et al., 1997); consequently the fluid can flow effectively. The specific surface area is large because of the rock size is small, which makes the fluid and particles of debris have an extensive contact, leading to an effective minerals extraction.

Cambrian carbonate is distributed sporadically in the South and Northwest. The purple to red clastic rocks of Jurassic and Cretaceous outcrop in small areas of the southern edge. The tertiary terrestrial clastic rocks are distributed mainly in the small, faulted basins of the Youjiang fault zone (Chen, 2002).

The Youjiang basin has a complex geological structure, experiencing many crustal movements. The early structure controlled and influenced the late structure; while the late structure inherited from and modified the early one. Especially, a series of folds and faults developed under the joint role of the North East stress of the India plate and the North West stress of the Pacific plate since the Middle-Late Triassic. The direction, intensity and feature of these structures vary in different areas of the basin. But overall, the deformation was controlled by the geological structure of the basin, forming three different structural features: uplift area, depression area and the contact zone of the two areas. The folds in the uplift area are wide and smooth with developed joints, rupture and crack. The folds in the depression area are commonly linearly closed (and even overturned), accompanied by thrust faults and the resultant faults. The contact zones between the uplift and depressed areas most likely developed faults, which cut through the Paleozoic continental crust and is known as the basement faults of the Triassic sedimentary cover. Most of them have the characteristics of inheritance and multiple activities (Guo et al., 1992; Luo, 1994). The tectonic lines in the basin are mainly in NW, NE and EW directions. In addition, the NW direction Youjiang fault zone, that traverses the entire Youjiang basin, has been an active basement fault cutting through the crust since the period of Indo-China - Yanshan, which might have controlled the general distribution of the gold in the basin (Tan, 1994).

The volcanic activities are frequent in the interior and edge of the Youjiang basin, occurring in all times with a climax in Late Permian except the middle-late Carboniferous. According to the research of Guangxi Bureau of Geology and Mineral Resources (1985) and Liu et al. (1993), the volcanic activities can be separated into two stages by Dongwu movement. The early (D_1-P_2) activities were generally in small scale as basic volcanic rocks of alkali and alkali calcic series, which were controlled mainly by the NW direction structures and distributed in the deep-water sediments of Guangnan-Napo basin and Longlin-Baise basin in the southern part of the Youjiang basin. The basalt has clear pillow and pore structures. The late (P_2-T_2) volcanic activities, having two climaxes in Late Permian and Early Triassic, were in large scale and great intensity. The Late Permian was represented by the famous basalt of Mount Emei, mainly distributing in northwest of the Youjiang basin and mostly in the form of continental eruption, including the basaltic lava and various types of basaltic volcanic clastic rocks, with a maximum thickness of 1600m. But the recent research (Kang et al., 2003) indicates that the submarine basalt eruption along the Napo fault and Youjiang fault in Northwest also occurred and cannot be ignored. The Triassic volcanic activity center drifted from the Napo-Pingxiang area of Early Trias to the Ningming-Chongzuo area of Middle Trias and the Shiwan mountains of Late Trias, and the lithology has also changed from basic rocks to medium-acidic rocks, belonging to Calc alkaline and alkaline series.

In general, the lithology of volcanic rocks in the Youjiang basin is characterized by the change from basic rocks and intermediate-basic rocks to intermediate-acidic rocks. The volcanic rocks also change from the NW direction to the NE direction, with the distribution center migrating southward. This may prove the evolution of the Youjiang basin from rift to foreland, but also reflect a fact that the Youjiang basin was controlled mainly by the Jinsha River-Red River fault zone of the southwest margin in its early history and then mainly by Qinzhou fault in the late period (in Middle-late Trias).

4. Geological and Geochemical Characteristics of Gold Deposits

The geological and geochemical characteristics of the Carlin-type gold deposits in the Youjiang basin (Hu et al., 2002) is very similar to that of the counterparts in the United States (Cline et al., 2005), though there are significant differences in ore-bearing strata and metallogenic epoch. The Carlin-type gold deposits of the Youjiang basin are mainly hosted in sedimentary rocks, but the epoch span of ore-bearing strata is very big, including the Cambrian, Devonian, Carboniferous, Lower Permian, Lower

Triassic, Middle Triassic and so on, with most deposits being hosted in the Middle Triassic (Zhu et al., 1998). The main ore-bearing rocks are argillaceous siltstones, silty mudstone and impure carbonate, and many of them are rich in carbonaceous matter (Liu et al., 1993; Gu et al., 2011). The details of representative Carlin-type gold deposits of the Youjiang basin can be found in Peters et al. (2007), Su et al. (2009), and Chen et al. (2011). The following is only an overview of the characteristics of these deposits.

The location of ore deposits is controlled by the sedimentary facies, and all the ore-containing layers or alteration zones are distributed over the petrographic boundary of the clastic rocks and carbonate rocks. That is, the ore deposits formed mostly in the microclastic rocks over the transition section of the outer or upper part of carbonate platform margin (Liu et al., 1993), refer to Figure 3.

The deposits are obviously controlled by the structures of faults and folds, and the ore bodies are affected by the secondary faults that are of high-angle beside the main faults and cut through the fold axis, or controlled by the interlayer faults nearby the unconformity surface (Yang et al., 1994; Chen, 2002).

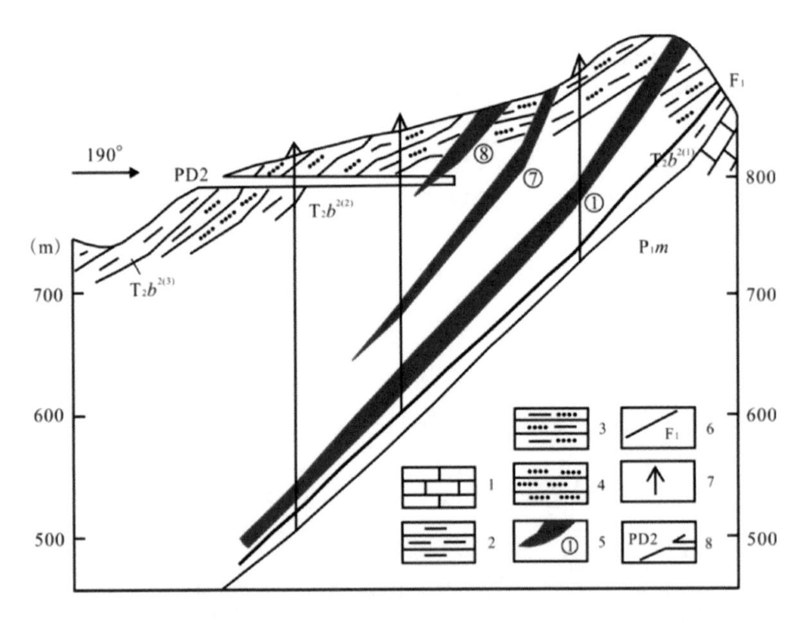

Figure 3. The cross section illustrating stratigraphic and structural control on orebodies. This is the 19th prospecting profile of the Mingshan gold deposit, northwestern Guangxi(Modified after Wang, 1998). 1. Limestone, 2. Mudstone, 3. Pelitic siltstone, 4. Siltstone, 5. Orebody, 6. Fault, 7. Drill hole, 8. Adit.

These structures mainly formed in the period of Indosinian-Yanshanian. Though there is no accurate chronological data for the metallogenic age of ore deposits, the ore-forming age is estimated to be between 193-82Ma (Zhang et al., 1993; Hu et al., 2002; Chen et al., 2007), which is consistent with the age of the ore-controlling structures.

The alteration-related mineralization mainly includes silicification, pyritization, argillization, carbonation and more. The associated ore rocks are mainly in disseminated or brecciated forms. The gold contained in pyrite, arsenopyrite and other sulfides is mainly in solid solution or microscopic-submicroscopic particles (nanometer particles) (Li et al., 1989; Guo, 1992; Yang et al., 1994; Su et al., 2009). Also, a very little of natural gold is present in the ore rocks (Wang et al., 1992; Su et al., 2008).

Fluid inclusions are very small, mostly in 3-10μm (Hu et al., 2002), and few can reach 20μm (Su et al., 2009). The ore-forming fluid system is the H_2O-NaCl-CO_2, with a low temperature (250 ±50℃) and a low salinity (<10 wt% NaCl) (Pang et al., 2001; Hu et al., 2002).

The majority of hydrogen-oxygen isotopes of the ore-forming quartz show that the ore-forming fluid is mainly the evolved atmospheric precipitation (Li et al., 1989; Guo, 1992; Wang et al., 1992; Hu et al., 2002; Pang et al., 2005). However, some deposits also show the characteristics of hydrogen-oxygen and argon isotopes with the involvement of magmatic water (Hu et al., 1995). The recent hydrogen-oxygen isotope data reveal that the metamorphic fluids were involved in the mineralization (Hofstra et al., 2005; Su et al., 2009). Comparing the sulfur isotope of pyrite in ore rock with that in unaltered host rock (Hu et al., 2002) shows a consistent isotopic range, which indicates that the sulfur in the ore rock was derived from the host rock strata. The carbon isotopic studies also indicate that the carbon in the ore rock was derived from carbonate strata (Liu et al., 1998). The silicon (Liu et al., 1998) and lead isotopes (Zhu et al., 1998) show that the ore-forming fluids were possibly from the deep-source magma. The high content of trace elements Co, Ni, Cu, Pb, Zn, and Pt in the fluid inclusions suggests that the minerals originated from the differentiated ore-forming fluids from the upper mantle (Su et al., 2001).

According to the above description, the Carlin-type gold deposits of the Youjiang basin exhibit similar geological, occurrence and mineralization characteristics. But their geochemical characteristics are very different, and sometimes contradictory even for the same deposit. A typical example is the Mingshan gold deposit in the northwest of Guangxi.

Figure 4. Ductile deformation of mineralized pyrite-quartz vein in the Mingshan gold deposit, northwest Guangxi.

Figure 5. Photomicrograph showing banded extinction and pressure dissolved cleavages in hydrothermal quartz in the ore from the Mingshan gold deposit, northwest Guangxi (Under transmitted cross-polarized light, the whole view is 3mm wide).

Figure 6. Photomicrograph showing NW direction, parallel behm lamellae in wavy extinction in hydrothermal quartz in the ore from the Mingshan gold deposit, northwest Guangxi (Under transmitted cross-polarized light, the whole view is 3mm wide).

The studies on elements H, O, S in quartz and static ^{36}Ar isotope in the ore-forming stage and the fluid inclusion data indicate that the ore-forming fluid originated from the mix of atmospheric precipitation, magmatic water and strata water (Hu et al., 1995). The petrochemistry of the quartz porphyry near the deposits also indicate that the origin of the gold was partly from the magmatic differentiation process, that served as the heat source for the activation and migration of ore-forming materials in the strata (Huang et al., 2001). The recent hydrogen-oxygen isotope data of fluid inclusions show the characteristics of metamorphic water (Chen, 2010). The ductile deformation of mineralized pyrite-quartz vein (Figure 4), microfabric of hydrothermal minerals (Figure 5, Figure 6), and chemical composition of minerals indicate that the deformation and metamorphism may have controlled the formation of deposits (Pang et al., 2012).

5. GENETIC MODELS OF GOLD DEPOSITS

There have been many debates on the formation of Carlin-type gold deposits. Most genetic models proposed in recent years have placed the deposits in the context of a long-term evolution of regional crust. For example, Cline et al. (2005) systematically compared the key geological features of the five major ore belts of northern Nevada, and linked the regional tectonic evolution with the mineral deposits based on the accumulated geochemical data; while Emsbo et al. (2006) linked other types of gold deposits of the northern Nevada together with the formation Carlin-type gold deposits.

Previous studies on the Carlin-type gold deposits of the Youjiang basin show that a few deposits indeed demonstrate a temporal link with magmatic rocks, such as the gold deposits related to the diabase in the period of Indo-China and Yanshanian (Xiao, 1997; Pang et al., 1998; Qin et al., 2003; Peters et al., 2007) and the gold deposits related to the quartz porphyry in the Yanshanian (Hu et al., 2002), although no magmatic rock outcrops are found around many deposits in the Yunnan-Guizhou-Guangxi regions. These deposits were produced in the fracture zones of rock mass and in the contact zones of rock mass and host rocks. Quartz porphyry outcrops are found nearby some Carlin-type gold deposits, such as Mingshan gold deposits in the northwest of Guangxi. Also, some ultrabasic rock veins, diabase, quartz porphyry and granite porphyry of Yanshanian are exposed in this region (Regional Geology of Guangxi, 1985; Regional Geology of Guizhou, 1987; Guo et al., 1992). The gravity and aeromagnetic data indicate that there is a large concealed rock mass at depth (Yang et al., 1994). It cannot be denied that atmospheric precipitation is one of the main characteristics of many fluid inclusions of Carlin-type deposits (Li et al., 1989; Guo et al., 1992; Hu et al., 2002). At least, atmospheric precipitation was involved in the late stage of ore formation and at shallow depth, regardless its concrete roles in the ore-forming processes. Indeed, atmospheric precipitation is present in many hydrothermal ore-forming fluids.

Though the metamorphic degree of the outcropped stratum in this region is low, the hydrogen-oxygen isotope data (Su et al., 2009a), structural analysis (Peters et al., 2007; Chen et al., 2011) and the microstructure and chemical composition of hydrothermal minerals (Pang et al., 2012) show that the deformation and metamorphism (occurring likely in deep crust) are the important factors for ore-forming processes.

Studies indicate that the involvement of organic fluids in ore formation is also very obvious in this region (Li et al., 1996; Zhang et al., 1999; Liu et al., 2002; Gu et al., 2011). The ore-forming fluids likely formed at the same time as oil and gas did, and migrated together with them, but located in different tectonic positions (Gu et al., 2011).

Some deposits have the characteristics of synchronal sedimentation and soft deformation, illustrating that they were mineralized in the diagenetic period of deposition (Liu et al., 2002). Also, the typical ore-containing bedded siliceous rocks represent an important symbol of sedimentary exhalative genesis (Liu et al., 1997).

Therefore, based on the recent geological and geochemical data, we speculate that the Carlin-type gold deposits in the Youjiang basin may not be the product of single-sourced fluids, but probably due to multi-sourced fluids involving multi-period and multi-stage. The ore formation involves various geological processes in relation to the basin evolution. Therefore, we establish the following ore-forming model of the Carlin-type gold deposits, which is adapted to the basin evolution. The model can be divided into sedimentary diagenetic stage, orogenic metamorphic stage, and postorogenic extension and magmatic activity stage.

Hercynian-early Indosinian Syndepositional Mineralization (D- End of T_2)

During this period, the Youjiang basin experienced the evolution from the continental marginal rift basin to the back-arc rift basin, and then to the (back-arc) foreland basin (Zeng et al., 1995; Du et al., 2009). Many basic volcanic rocks erupted in different times, and the submarine volcanic eruption formed the widely-distributed basic basalt and basaltic volcanic clastic rocks with a high Au abundance of 23.8-200ppb (Guizhou Bureau of Geology and Mineral Resources, 1997; Li et al., He, 1992), which formed the main source of gold in the basin strata by sea solution. Sometimes the submarine volcanic eruption may also have formed the gold deposits of jet deposition type (Liu et al., 1997). The gold brought by basic magma from depth was absorbed mainly by clay and organic matters, precipitating therein and leading to the late ore formation.

Li et al. (1996) show that the average gold abundance of strata in this region is not high. The majority is lower than the Clark value of crust with some slightly higher than that value, and only few strata have a high gold abundance. Comparing the gold content of the strata of the same layer in different areas indicates that the gold abundance is irrelevant to the layer but is controlled mainly by the sedimentary facies, rock facies and lithology. Specifically, the gold abundance of the slope facies linking the carbonate platform and the clastic basin facies is high. The clastic rocks have a highest gold abundance, followed by the impure carbonate, and the pure carbonate has a lowest value. The gold abundance in clastic rocks decreases from fine sandstone, siltstone, mudstone to shale; but when mudstone and shale contain more organic carbon and pyrite, the gold abundance increases greatly. The gold abundance in carbonate decreases from dolomitic limestone, bioclastic limestone to limestone. According to the above distribution of gold abundance, it seems that gold tends to track the porosity and organic matters. This explains the effect of fluid on the enrichment and transport of gold in the process of diagenetic compaction.

The gold contents in some strata are 5-10 times higher than the regional background value (Li et al., 1989) and constitute the important ore source strata of Carlin-type gold deposits in the basin, such as the Yilan unit of Lower Devonian, Da-Chang stratum on the top of Maokou unit of Lower Permian, the bottom of Longtan unit of Upper Permian, Xinyuan unit (southwest of Guizhou) / Baipeng unit (northwest of Guangxi) of Middle Trias. The Triassic clastic turbidite is of large thickness, and has developed synsedimentary faults in the transition zone of itself and carbonate platform. The synsedimentary faults not only served as good channels for the fluids migration, but also acted as an accumulation destination of the fluids discharged from the depth of the basin, thus represent a favorable locale of syngenetic - quasi syngenetic ore formation. When the basinal fluids migrated to the high-elevation areas along the contemporaneous faults, they might discharge to the seafloor, unloading as a submarine hot water deposition (being diluted at different degrees by normal sediments accumulated at the same time), thereby generating the mineralization type with syndepositional mineralization characteristics nearby the platform. The basinal fluids might also unload the minerals directly in the contemporaneous fault system (hot water channel system) and generate the mineralization type with epigenetic mineralization characteristics far away from the platform (Liu et al., 2002).

Metamorphism and Mineralization of Indosinian Folding Orogeny Period (End of T_2-J_2)

From the end of T_2 to J_2, the Youjiang basin was affected mainly by the northeast subduction of Indosinian plate, which caused strong fault-fold zones in NW direction in the interior of the basin, especially evident in the large gold mines (Chen et al., 2011). This subduction also activated previously existing NW direction synsedimentary faults and formed some secondary reverse faults in same direction in its either sides, which controlled the location of gold deposits. In fact, many NW direction reverse faults represent one of the main ore-controlling structures, such as in the Lannigou, Shuiyindong, Jinya, Gaolong and Minghsan gold deposits. The ductile shear effect in this period was obvious, and the ductile deformation can be identified before and in the process of ore formation, such as the ductile deformation of mineralized quartz pyrite veins, the Behm lamellae, wavy extinction, zonal extinction, pressure solution cleavage and slip dislocation etc. (Pang, 2012). The metallogenic materials of metamorphic mineralization of this period were from the activation of mineral elements in the host rocks. The pre-concentration of gold caused by the sedimentary and diagenetic processes of the host rocks laid an important material foundation for the late ore formation. Due to the strong influence and transformation of the ore metallogeny in the late period, the ore formation in this period is difficult to be identified. Recently, the hydrogen-oxygen isotopes of fluid inclusions (Hofstra et al., 2005; Su et al., 2009; Chen, 2010), structural analysis (Peters et al., 2007; Chen et al., 2011), hydrothermal mineral microstructure and mineral chemical composition (Pang et al., 2012), and the isotope chronology (Chen et al., 2007) have collectively indicated that the metamorphic mineralization in Indosinian period is very important for ore formation. Gu et al. (2011) conducted a research on the metal deposits and the oil and gas of the Youjiang basin, and concluded that they have the characteristics of symbiotic differentiation, and that the Indosinian folding orogeny from south to north drove the basinal fluids to flow from south to north under the effect of tectonic stress and gravity. This process, on the one hand, resulted in the change of the original oil and gas distribution, and oil and gas further migrated and accumulated towards the structural high locations in the basin; On the other hand, it created a spatial zonation of the oil and gas and the metal deposits distributing from north (South East) to south (North West), and from orogenic belt to foreland basin.

Magmatic Activities and Atmospheric Precipitation Infiltration Mineralization in Yanshanian (J_3-K_2)

Although the Circum Pacific tectonic domain began to be active during the Indosinian period, it had a less effect than the Tethys tectonic domain did on the Youjiang basin. However, the Circum Pacific tectonic reached its peak in Yanshanian, since many large-scale magmatic activities and metallogenesis occurred in this period in eastern China (Hua et al., 1999). Due to the remote effect of the westward subduction of the Pacific plate, the reverse faults formed before Indosinian rift reactivated, and the folds deformed or fractured, leading to the conditions for a deep infiltration of atmospheric water. This facilitated the decompression melting of deep rocks, and formed quartz porphyry, granite porphyry and so on. Some lamprophyre and alkaline ultrabasic rocks also outcrop in the basin (Chen et al., 2012).

The NW-direction ore-controlling reverse faults are cut through by the NE-direction faults in many deposits, and this is proved by the mineralized quartz pyrite veins being cut off by the quartz of late periods (Pang et al., 2012). Although the mineralization age of gold deposits of Youjiang basin is still inconsistent, most studies indicate that it was in the Middle-late Yanshanian, corresponding to the extensional-collapse and lithospheric-thinning stage (135~100Ma) and the weak-compressive deformation stage (100~83Ma) after the strong-compressive intracontinental orogenic stage (165±5Ma~136Ma) (Dong, 2007). During this period of time, the mantle upwelling caused the crust thinning and magmatic activities in Youjiang basin, resulting in the formation of the ultrabasic rocks and acid rocks. A part of the gold was probably supplied by the magmatism, but the main role of the magmatic activities was to provide the heat source, thus accelerated the infiltration of atmospheric precipitation. In particular, the extension after strong compression may form a large number of fractures, making it easier for the atmospheric precipitation to leach gold from strata. That is probably why a lot of ore-forming fluids have the characteristics of atmospheric precipitation (Hu et al., 2002), with some involving magma fluids (Zhu et al., 1998; Liu et al., 1999; Su et al., 2001). Therefore, the Yanshanian may be the period of main gold ore formation in this study region.

ACKNOWLEDGMENTS

This research was supported by the National Natural Science Foundation of China through a grant (41362006) and the Guangxi Natural Science Foundation through a grant (2013GXNSFAA019275) to Baocheng Pang. This was also partly supported by the Guangxi Natural Science Foundation through a research grant (2012GXNSFAA053180) to Jianwen Yang. The work was sponsored by the *Program to Sponsor Teams for Innovation in the Construction of Talent Highlands in Guangxi Institutions of Higher Learning.*

REFERENCES

Chen, K. L., 2002. Geology of gold deposits in Guangxi. *Nanning: Guangxi Scientific and Technical Publishing House*, pp. 192-295 (in Chinese).

Chen, M. H., 2010. Geological and ore-forming fluid characteristics of the Mingshan gold deposit in western Guangxi. *Mineral Deposits*, 29(supl):913-914(in Chinese).

Chen, M. H., Lu, G,, Li, X. H., 2012. Muscovite 40Ar/39Ar Dating of the Quartz Porphyry Veins from Northwest Guangxi, China, and its Geological Significance. *Geological Journal of China Universities*, 18(1):106-116(in Chinese with English abstract).

Chen, M. H., Mao, J. W., Qu, W. J., Wu, L. L., Uttley, P. J., Norman, T., Zheng, J. M., Qin, Y. Z., 2007. Re–Os dating of arsenian pyrites from the Lannigou gold deposit, Zhenfeng, Guizhou Province, and its geological significances. *Geol. Rev.*,53:371–382 (in Chinese with English abstract)

Chen, M. H., Mao, J. W., Bierlein, F. P., et al., 2011. Structural features and metallogenesis of the Carlin-type Jinfeng (Lannigou) gold deposit, Guizhou Province, China. *Ore Geology Reviews*, 43: 217-234(in Chinese with English abstract).

Chen, M. H., Huang, Q. W., Hu, Y., Chen, Z. Y., Zhang, W., 2009. Genetic types of phyllosilicate (mica) and its 39Ar–40Ar dating in Lannigou gold deposit, Guizhou province. China. *Acta Mineralogica Sinica*, 29:353– 362 (in Chinese with English abstract).

Cline, J. S., Hofstra, A. H, Munteau, J. L, Tosdal, R. M., and Hickey, K. A., 2005. Carlin-type gold deposits in Nevada: Critical geologic characteristics and viable models. *Economic Geology 100th Anniversary Volume*, 451-484.

Cunningham, C. G., Ashley, R. P., Chou, I. M., Huang, Z. H.,Wan, C. Y., Li, W. K., 1988. Newly discovered sedimentary rock-hosted disseminated gold deposits in the People's Republic of China. *Economic Geology*, 83:462–469.

Dong, S. W., Zhang, Y. Q., Long, C. X., Yang, Z. Y., Ji, Q., Wang, T., Hu, J. M., Chen, X. H., 2007, Jurassic Tectonic Revolution in China and New Interpretation of the Yanshan Movement. *Acta Geologica Sinica*, 81(11): 1449-1461(in Chinese with English abstract).

Du, Y. S., Huang, H. W., Huang, Z. Q., Xu, Y. J., Yang, J. H., Huang, H., 2009, Basin Translation from Late Palaeozoic to Triassic of Youjiang Basin and Its Tectonic Significance.*Geological Science and Technology Information*, 28(6):10-15(in Chinese with English abstract).

Emsbo, P., Groves, D. I., Hofstra, A. H., Bierlein, F. P., 2006, The giant Carlin gold province: a protracted interplay of orogenic, basinal, and hydrothermal processes above a lithospheric boundary. *Mineralium Deposita*, 41:517–525.

Gu, X. X., Zhang, Y. M., Li, B. H., et al., 2011. Hydrocarbon- and ore-bearing basinal fluids: a possible link between gold mineralization and hydrocarbon accumulation in the Youjiang basin, South China. *Mineralium Deposita*, DOI 10.1007/s00126-011-0388-x.

Guangxi bureau of geology, 1985. Regional geology of Guangxi. *Geology publishing house, Beijing* (in Chinese with English abstract).

Guizhou bureau of geology, 1987. Regional geology of Guizhou. *Geology publishing house, Beijing* (in Chinese with English abstract).

Guo, J. H., Huang, D. B., Shi, L. D., et al., 1992, Micro-granular gold deposits in northwest of Guangxi and models for ore-forming and ore prospecting. S*eism publishing house, Beijing*, (in Chinese).

He, S. L., 1992. Preliminary explanation of Au formation in southwestern Guizhou by using geological and geoehemieal data. *Geology of Guizhou*, 9(2):150-160(in Chinese with English abstract).

Hofstra, A. H., Zhang, X. C., Emsbo, P., et al., 2005. Source of ore fluids in Carlin-type gold deposits in the Dian-Qian-Gui area and West Qinling belt, P.R. China: Implications for genetic models, *in* Mao, J. W., and Bierlein, F. P., eds., Mineral deposits research: Meeting the global challenge: Heidelberg, *Springer-Verlag*, 1, 533-536.

Hu, R. Z., Su, W. C., Bi, X. W., 1995. A possible evolution way of ore-forming hydrothermal fluid for the Carlin-type gold deposits in the Yunnan-Guizhou-Guangxi Triangle area. *Acta Mineralogical Sinica*, 15 (2): 144-149 (in Chinese with English abstract).

Hu, R. Z., Su, W. C., Bi, X. W., Li, Z. Q., 1995. A possible evolutionway of ore-forming hydrothermal fluid for the Carlin-type gold deposits in Yunnan–Guizhou–Guangxi triangle area. *Acta Mineralogica Sinica*, 15(2):144–149(in Chinese with English abstract).

Hu, R. Z., Su, W. C., Bi, X. W., et al., 2002. Geology and geochemistry of Carlin-type gold deposits in China. *Mineralium Deposita*, 37: 378-392.

Hua, R. M., Mao, J. W., 1999. A preliminary discussion on the Mesozoic metallogenic explosion in east china. *Mineral Deposits*,18(4): 300-307.

Huang, Y. Q., Cui, Y. Q., 2001. The relationship between magmatic rocks and gold mineralization of Mingshan gold deposit of Lingyun, Guangxi. *Guangxi Geology*, 14(4):22-28(in Chinese with English abstract).

Kang Yunji, Zhang Geng, Cai Heqing, 2003. Geochemical characteristics of magmatic rocks in Youjiang basin. *South Land Resources*, 25,(8):24-27 (in Chinese).

Li, W. K., Jiang, X. S., Jiu, R. H., et al., 1989. Geological characteristics and mineralization of micro-disseminated gold deposits in southwestern Guizhou.province. In: Collected works on regional ore-forming situation of main gold deposits in China (The area of soutwestern Guizhou). *Geology publishing house, Beijing*, 1-81 (in Chinese).

Li, Z. P. and Peters, S. G., 1998. Comparative Geology and Geochemistry of Sedimentary- Rock-Hosted (Carlin-Type) Gold Rock-Hosted (Carlin-Type) Gold Deposits in the People's Republic of China and in Nevada, USA. U.S. *Geological Survey Open-File Report*, 98-466.

Li, Z., Liu, T. B., 1996. On basin controlling micro-disseminated gold deposits in Youjiang basin, south Chian. *Geology and Geochemistry*,(1)47-51(in Chinese with English abstract).

Liu, B. J., Xu, X. S., Pan, X. N., et al., 1993. Evolution and ore-forming of sedimentary paleocontinental crust in southern China. *Science publishing house, Beijing* (in Chinese).

Liu, J. J., Liu, J. M., Gu, X. X., Lin, L., 1997. Sedimentary exhalative origin of micro-disseminated gold deposits in southwestern Guizhou. *Chinese Science Bulletin*, 42:2126-2127 (in Chinese).

Liu, J. M., Liu, J. J., Gu, X. X., 1997. Basinal fluids and their related ore deposits. *Acta Petrologica et Mineralogica*, 16(4): 341-352 (Chinese with English abastract).

Liu, J. M., Liu, J. J., Zheng, M. H., Gu X. X., 1998. Stable isotope compositions of micro-disseminated gold and genetic discussion. *Geochemica*, 27(6):585-591(in Chinese with English abstract).

Liu, W. J., Zeng, Y. F., Zhang, J. Q., Chen, H. D., Zheng, R. C., 1993. The Geochemical characterisitics of volcanic rocks and structural setting in Youjiang basin. *Geology of Guangxi,* 6(2):1-14(in Chinese with English abstract).

Liu, X. F., Su, W. C., Zhu, L. M., 1999. An approach on mechanism of juvenile fluid mineralization for Carlin-type gold deposits in Yunnan-Guizhou-Guangxi. *Geology and prospecting*, 35(1): 14-19 (in Chinese with English abstract).

Liu, X. F., Ni, S. J., Su, W. C., 1996. characteristics of isotope geochemistry and plutonic-origin fluid mineralization for Carlin-type gold deposits in the Yunnan-Guizhou-Guangxi. *Journal of Mineralogy and Petrology*, 16(4):106-111(in Chinese with English abstract).

Liu, X. F., Ni, S. J., Lu, Q. X., Jin, J. F., Zhu, L. M., 1998. The trace of silicon isotope geochemistry of mineralization materials-origin for Carlin-type gold deposits – A case study of the gold deposits in southwestern Guizhou and Northwestern Guangxi Province. *Gold Science and Technology*, 6(2):18-26(in Chinese with English abstract).

Liu, J. M., Ye, J., Ying, H. L., et al., 2002. Sediment-hosted micro-disseminated gold mineralization constrained by basin paleo-topographic highs in the Youjiang basin, South China. *Journal of Asian Earth Sciences*, 20:517-533.

Luo, X. H., 1994. Geological characteristics, forming mechanism and prospect on Lannigou gold deposit in Zhengfeng County, Guizhou Province. In: Liu, D. S., Tan, Y. J., Wang, J. Y., Jiang, S. F. (Eds.), Chinese Carlin-type Gold Deposits. *University of Nanjing Press, Nanjing*, pp. 100–115 (in Chinese).

Luo, X. H., 1994. Geological characteristics, forming mechanism and prospect on Lannigou gold deposit in Zhengfeng County, Guizhou Province. In: Liu, D. S., Tan, Y. J., Wang, J. Y., Jiang, S. F. (Eds.), Chinese Carlin-type Gold Deposits. *University of Nanjing Press, Nanjing*, pp. 100–115 (in Chinese).

Muntean, J. L., Cline, J. S., Simon, A. C., et al., 2011. Magmatic–hydrothermal origin of Nevada's Carlin-type gold deposits. *Nature Geoscience*, 4:122-127.

Pan, J. Y., Zhang, Q., Shao, S. X., 1998. A new type of fine disseminated gold deposits found in Northwest Guangxi, China, *Gold*, 19(7):3-6(in Chinese with English abstract).

Pang, B. C., Lin, C. S., 2001. Discussion on genesis of microgranular gold deposits in youjiang basin. *Geology and prospecting*, 37(4): 9-13 (in Chinese with English abstract).

Pang, B. C., Xiao, H., Chen, H. Y., Yang, X. F., Fu, W., Wang, B. H., Li, W. L., 2012. Metamorphic ore-forming process of the Mingshan gold deposit in Guangxi: evidence from microfabric and composition of pyrite. *Mineral Deposits*, 31(supl):745-746 (in Chinese).

Pang B. C., Lin, C. S., Luo, X. R., Hu, C. Y., Zhuang, X. G., 2005. The characteristic and origin of ore-forming fliud from micro-disseminated gold deposits in Youjiang basin. *Geology and Prospecting*, 41(1):13-17(in Chinese with English abstract).

Peters, S. G., Huang, J. Z., Li, Z. P., et al., 2007. Sedimentary rock-hosted Au deposits of the Dian-Qian-Gui area, Guizhou, and Yunnan provinces, and Guangxi district, China. *Ore Geology Reviews*, 31: 170–204.

Qin, J. H., Wu, Y. L., Yan, Y. J., Zu, Z. F., 1996. Hercynian-Indosinian sedimentary-tectonic evolution of the Nanpanjiang basin, *Acta Geologica Sinica*, 70(2):99-107(in Chinese with English abstract).

Qin, W. M., He, Z. M., 2003. Geological features of gold deposits that have relationship with diabase in the northwest of Guangxi : Taking example of the Longchuan gold deposit in Baise, *Gold Geology*, 9(3):49-54(in Chinese with English abstract).

Shui, T., 1987. Tectonic situation of continental basement in southeastern China. *Science in China (series B)*, (4): 414-422 (in Chinese with English abstract).

Su, W. C., Hu, R. Z., Yang, K. Y., 1998. Chronology of fluid inclusions in Carlin-type gold deposits in southwestern China as exemplified by Lannigou and Yata gold deposits in Guizhou province. *Acta Mineralogical Sinica*, 18(3): 359-362 (in Chinese with English abstract).

Su, W. C., Hu, R. Z., Qi, L., Fang, W. X.,2001. Trace elements in fluid inclusions in the Carlin-type gold deposits, southwestern Guizhou province. *Geochemica*, 30(6):512-516(in Chinese with English abstract).

Su, W. C., Heinrich, C. A., Zhang, X. C., et al., 2009a. Sediment-Hosted Gold Deposits in Guizhou, China: Products of Wall-Rock Sulfidation by Deep Crustal Fluids. *Economic Geology*, 104: 73-93.

Su, W. C., Hu, R. Z., Xia, B., et al., 2009b. Calcite Sm-Nd isochron age of the Shuiyindong Carlin-type gold deposit, Guizhou, China. *Chemical Geology*, 258: 269-274.

Su, W. C., Xia, B., Zhang, H. T., Zhang, X. C., and Hu, R. Z., 2008. Visible gold in arsenian pyrite at the Shuiyindong Carlin-type gold deposit, Guizhou, China: Implications for the environment and processes of ore formation. *Ore Geology Reviews*, 33:667-679.

Tan, Y. J., 1994. Geology of Carlin-type gold deposits in the Dian– Qian–Gui area. In: Liu, D. S., Tan, Y. J., Wang, J. Y., Jiang, S. F. (Eds.), Chinese Carlin-type Gold Deposits. *University of Nanjing Press, Nanjing*, pp. 142-159 (in Chinese).

Wang, D. H., Lin, W. W., Yang, J. M., Yan, S. H., 1999. Controlling effects of the mantle plume on the Jiaodong and Dian-Qian-Gui gold_concentration areas. *Acta Geoscientia Sinica*, 20(2):157-162(in Chinese with English abstract).

Wang, G. T., 1989. Preliminary discussion on the ore-forming mechanism of the micro-grained disseminated JY gold deposit in northwestern Guangxi. *Guangxi Geology*, 2(2): 15-24 (in Chinese with English abstract).

Wang, G. T., 1992. Three strontium–rubidium isochron ages in northwest Guangxi. *Geology of Guangxi*, 5(1):29-35 (in Chinese).

Wang, H. Z. (chief editor), 1985. Paleogeography atlas of China. Beijing: *Atlas Publishing House* (in Chinese).

Wang, L. D., 1998. Geological characteristics and indicaters of Mingshan microfine- grained disseminated type gold deposit, Guangxi. *Guangxi Geology*, 11(3):21-25 (in Chinese).

Wang, K. R., Zhou, Y. Q., Li, F. Q., Sun, L. G., Wang, J. X., Ren, C. G., Zhou, S. J., Tang, J. Y., Yang, F. J., 1992. SPM and SEM study on the occurrence of micrograined in the Jinya gold deposit, Guangxi. *Chinese Science Bulletin*, 9:832-835 (in Chinese).

Wang, X. Z., Cheng, J. P., Zhang, B. G., et al., 1992. Geochemistry of transformed gold deposits in China. *Science publishing house, Beijing*(in Chinese).

Wang, Y. G., Sue, S. T., Zhang, M. F., 1994. Structure and Carlin-type gold deposits in southwestern Guizhou Province. *Geological Publishing House, Guiyan*. (in Chinese).

Wu, Y. L. Qin, J. H., Zhu, Z. F., Mu, C. L., Luo, C. X., Tan, Q. Y., 1990. Evolution of a Hercynian-Indosinian foreland basin on the southeastern margin of the Yangzi plate and tectonics in south china. *Lithofacies Paleography*, 2:8-15 (in Chinese).

Xiao, L., 1997. New type of micro-disseminated gold deposit: the characteristcs and ore finding sign of micro-disseminated gold deposit exposed in diabase. *Geology and prospecting*, 33(6):1-6(in Chinese with English abstract).

Yang, K. Y., Zhang, Z. R., Chen, F., et al., 1994. Metallogenic characteristics and distribution rules of Carlin-type gold deposits in Yunnan-Guizhou-Guangxi area. In: China acadamy of sciences eds. The new progress for the researchment of Chinese gold deposits (continue part). *Seism. publishing house, Beijing*, pp 284-291(in Chinese).

Zeng, Y. F., Liu, W. J., Chen, H. D., et al., 1995. Tectonic and sedimentary evolution of Youjiang complex basin in south of china. *Acta Geologic Sinica*, 69(2): 113-124 (in Chinese with English abstract).

Zhang, F., Yang, K. Y., 1992. The study on fission-track dating of metallogenic epoch for microdisseminated gold deposits in the southwestern Guizhou. *Chinese Science Bulletin*, 37(7): 1593-1595 (in Chinese with English abstract).

Zhang, J. R., Lu, J. J., Zhang, X. H., et al., 1997. Physical characteristics of triassic sedimentary rocks and its relation to gold mineralization in

Youjiang rift zone. *Mineral Deposits*, 16(4): 340-348 (in Chinese with English abstract).

Zhang, Z. J., Zhang, W. H., 1999. Investigation into metallogenic fluid feature in Carlin-type gold deposits and its relation to mineralization in southwest Guizhou province. *Earth Science*, 24(1): 74-78(in Chinese with English abstract).

Zhang, Z. J., Zhang, W. H., 1999. Investigation into metallogenic fluid feature in Carlin-type gold deposits and its relation to mineralization in southwest Guizhou province. *Earth Science*, 24(1): 74-78(in Chinese with English abstract).

Zhu, L. M., Liu, X. F., Jin, J. F., He, M. Y., 1998. The study of the time-space distribution and source of ore-forming fluid for the fine-disseminated gold deposits in the Yunnan-Guizhou-Guangxi area. *Scientia Geologica Sinica*, 33(4):463-474(in Chinese with English abstract).

Zhu, Z. F., et al., 1992. Geotectonic and Indosinian movement in southern China. *Corpus of Lithofacies and Paleogeography* (8):1-7 (in Chinese).

In: Basins
Editor: Jianwen Yang

ISBN: 978-1-63117-510-7

Chapter 2

Numerical Simulation on Sedimentary Processes of Turbidity Current and Its Applications on Reservoir Prediction in the Qiongdongnan Basin, Northern South China Sea

Tao Jiang*[*], *Sulin Tang, Xinong Xie and Bo Wang
Key Laboratory of Tectonics and Petroleum Resources of Ministry of Education, China University of Geosciences, Wuhan, 430074, China

Abstract

The hydrocarbon explorations in deepwater areas around the world have achieved great success in past decades. The main targets of deepwater exploration are submarine fans. But some of them comprise shale or mud and therefore cannot be hydrocarbon reservoirs. Therefore, the reservoir prediction before drilling is essential and important for hydrocarbon exploration in deepwater areas. Unfortunately, the wells are few in deepwater areas, so they cannot be used to make effective predictions by conventional log-constrained inversion. Hydrodynamic simulation can illuminate the sedimentary processes of gravity flows,

[*] Corresponding author: Dr. Tao Jiang, Key Laboratory of Tectonics and Petroleum Resources of Ministry of Education, China University of Geosciences, Wuhan, 430074, China, Tel:(86)27-67883603 Fax: (86) 27-67883051, Email: taojiang@cug.edu.cn.

including turbidite currents, which are the main geneses of submarine fans. Combined with the geological background in the Qiongdongnan basin (QDNB), northern South China Sea, this study simulates the sedimentary processes and geometric shapes of turbidites composed of various size particles. Then, compared with seismic reflection features, the main component of the submarine fans interpreted in seismic data is estimated. The prediction has been confirmed by several recent drillings in the QDNB. Moreover, the results show that the slope gradient controls the development of turbidity currents and 1.5°-3° are more appropriate for triggering turbidity current that can flow a long distance along the slope. The turbidities are usually deposited as slope fans and basin floor fans, and the single turbidite has the thickest segment in the slope foot and thins toward the basin. We conducted the simulation with various grain sizes and observed the geometric shapes of formed submarine fans, and found that the coarse grain sizes generate thick turbidities with a limited spread, thicker and smaller, and vice versa.

Keywords: Turbidity current, Hydrodynamic simulation, Reservoir prediction, Qiongdongnan basin, South China Sea

1. INTRODUCTION

Many of large oilfields have been discovered in deepwater areas around the world, such as offshore West Africa and Gulf of Mexico, in the past decades, and their reservoirs are mainly turbidity current deposits due to their good sorting and sandy components (Pettingill and Weimer, 2002). However, not all turbidities are sandy and can be viewed as good hydrocarbon reservoirs. Some of submarine fans interpreted in seismic data may comprise other gravity flow deposits, such as debrites, slumps, mass transport complexes and so on (Pratson et al., 2000; Amy et al., 2005). These gravity flow deposits may not be good hydrocarbon reservoirs due to their low porosity and permeability. Therefore, reservoir prediction before drilling is important and essential for hydrocarbon exploration, especially in deepwater areas due to the high risk and expensive drilling fees. Traditionally, the log-constrained inversion is used to predict reservoirs in land or shallow water areas because a large amount of wells can be drilled. But usually, there are very few wells available in deepwater areas.

Recently, increased hydrocarbon exploration activities and successes in deepwater environments have driven the study of subsurface turbidity systems (Posamentier and Kolla, 2003; Ericilla et al., 2008). Events and mechanisms governing the initiation of turbidity currents have been illuminated with experimental observations and findings from field studies regarding the internal velocity and density structure (Meiburg and Kneller, 2010), which makes it possible to simulate the sedimentary processes and to predict their lithology under specific geological settings.

On the basis of the interpretation of seismic data and the integrated investigation in the Qiongdongnan basin (QDNB) of the northern South China Sea, some submarine fans are identified and the geologic models, including the slope gradient, slope length and provenance characteristics are built up. According to the principles of hydrodynamic simulation, the mathematic model can be obtained. Given the initial velocity, rough bed and boundary conditions, the forward simulations of turbidity sedimentary processes are executed with various grain sizes. The geometries of the formed turbidites are compared with those in seismic data, assuming the complex of recurrent turbidity currents under the same or similar geological environments. We change the grain size and repeat the forward simulations until a similar geometry coincides with that in seismic data. Several recent drillings in this area have confirmed the prediction methodology. Therefore, this study would be helpful to better understand the sedimentary processes of turbidity currents and reducing the risks associated with the future exploration efforts in this area and other similar deepwater areas.

2. GEOLOGICAL SETTING AND DEEPWATER HYDROCARBON EXPLORATION IN QIONGDONGNAN BASIN

A series of Cenozoic extensional basins are formed on the continental margin of the South China Sea. The QDNB lies in the northwestern part of the South China Sea between 108°52'E-110°47' and 16°47'N-19°00'N. Covering about 45,000 km^2, it trends east to northeast. To the west, the basin is bound by the Red River Fault and the Yinggehai Basin, to the east by the Pearl River Mouth Basin, and to the south by the Xisha Rise (Figure 1). The formation and evolution of QDNB are closely related to the opening of the South China Sea (Chen et al., 1993). Detailed studies on sedimentology, tectonics, and evolution of QDNB have been documented by several literatures (Hao et al., 2000; Xie et al., 2008; Yuan et al., 2009). There are two mega sequences: the rifting period (53.5-23.3 Ma) and the post-rifting period (23.3 Ma to Quaternary). The shelf-slope system has developed since the middle Miocene due to the anomalous post-rift subsidence and tectonic reactivation along the northern South China Sea margin, which controls the overall seafloor topography. The depositional environment in the study area is initially transformed from lacustrine to marine, and later from neritic to bathyal, starting in Eocene and continuing into the present (Xie et al., 2006). During the post-rifting stage, gravity flow deposits are developed, appearing as submarine fans and channels in high-resolution seismic data (Su et al., 2011).

A few deepwater hydrocarbon reservoirs such as Ya 13-1 and LW 3-1 gas fields were found in the northern South China Sea, and more attention has been paid to the deepwater slope area of the QDNB (Pang et al., 2006; Wu et al., 2008). The China National Offshore Oil Corporation (CNOOC) has acquired plenty of high-resolution seismic profiles in the past several years, covering the whole area of QDNB. Based on these seismic data, many submarine fans have been identified and some of them have been drilled as hydrocarbon targets.

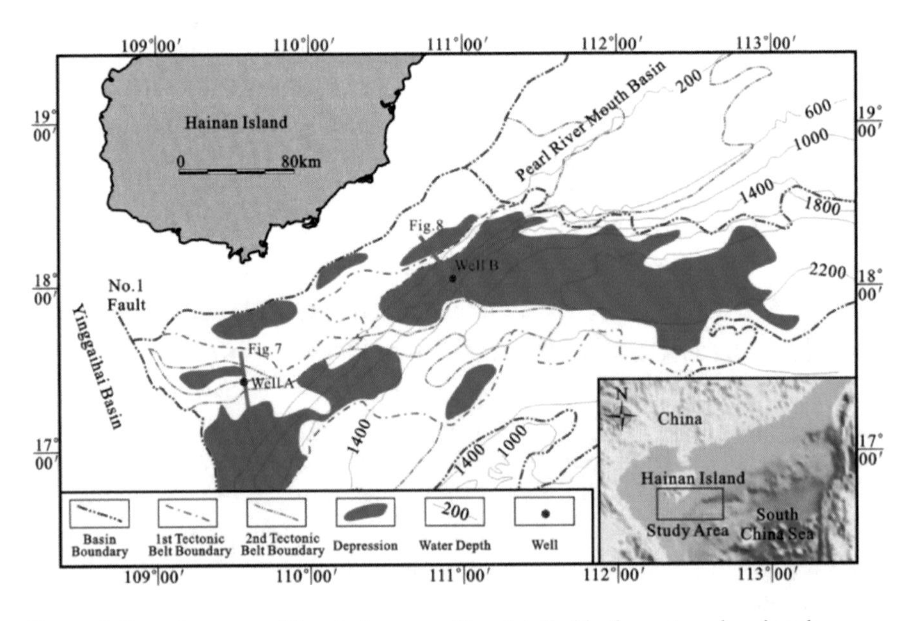

Figure 1. Location map of the study area with a detailed bathymetry, showing the major tectonics units, locations of seismic profiles and wells referred to in the context. Inset shows location of study area in South China Sea.

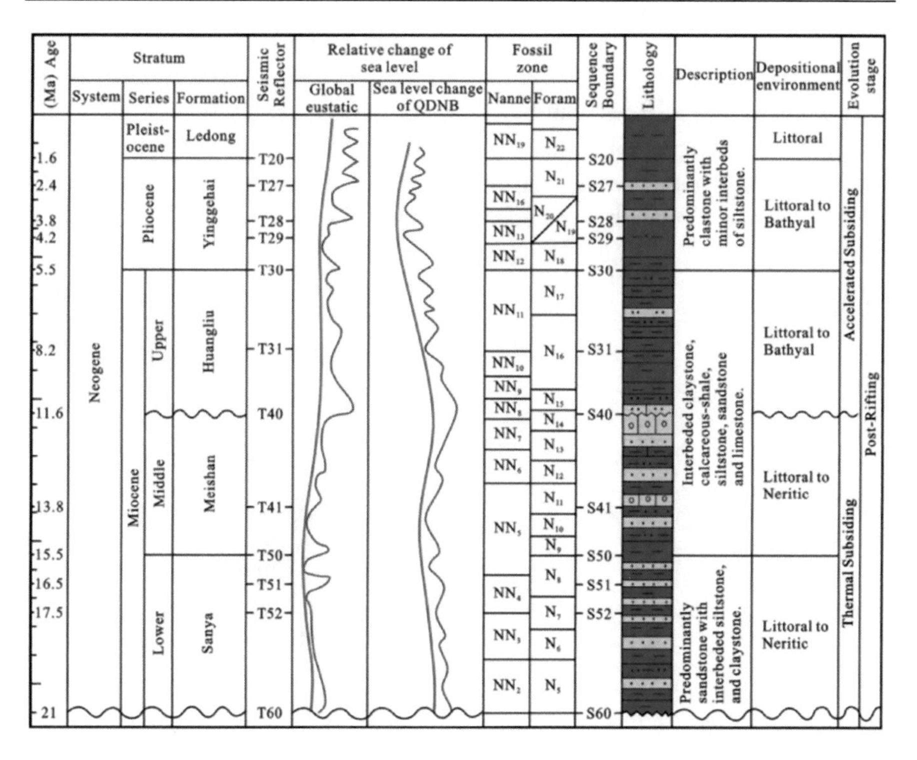

Figure 2. Lithologic framework, sedimentary environment, tectonic evolution and sea-level changes of the Miocene, Pliocene and Early Pleistocene in the QDNB. Since Middle Miocene, the shelf-slope system has been developed in the basin and amounts of submarine fans (lowstand fans) are deposited on the slope and basin floor fan.

3. Model Description

3.1. General Information

Mathematical and numerical models, when properly designed and tested against field or laboratory data, can provide significant knowledge of turbidity current dynamics as well as erosional and depositional characteristics. To date, there are various numerical studies dealing with turbidity current dynamics and the flow characteristics (Hartel et al., 2000; Kassem and Imran, 2001; Heimsund et al. 2002; Lavelli et al., 2002; Necker et al., 2002; Bombardelli et al., 2004; Imran et al., 2004; Banchette et al., 2005; Jiang et al., 2007; Mehdizadeh et al., 2008; Cantero et al., 2008; Singh, 2008).

Turbidity current flows can be considered multiphase systems since they consist of a primary fluid phase (water) and secondary granular phases (suspended sediment classes) dispersed into the primary phase. Therefore, turbidity currents can be modeled through the application of suitable multiphase numerical methods. In this study, the Computational Fluid Dynamics (CFD) method through the commercial software FLUENT was used to simulate the sedimentary processes of turbidity currents.

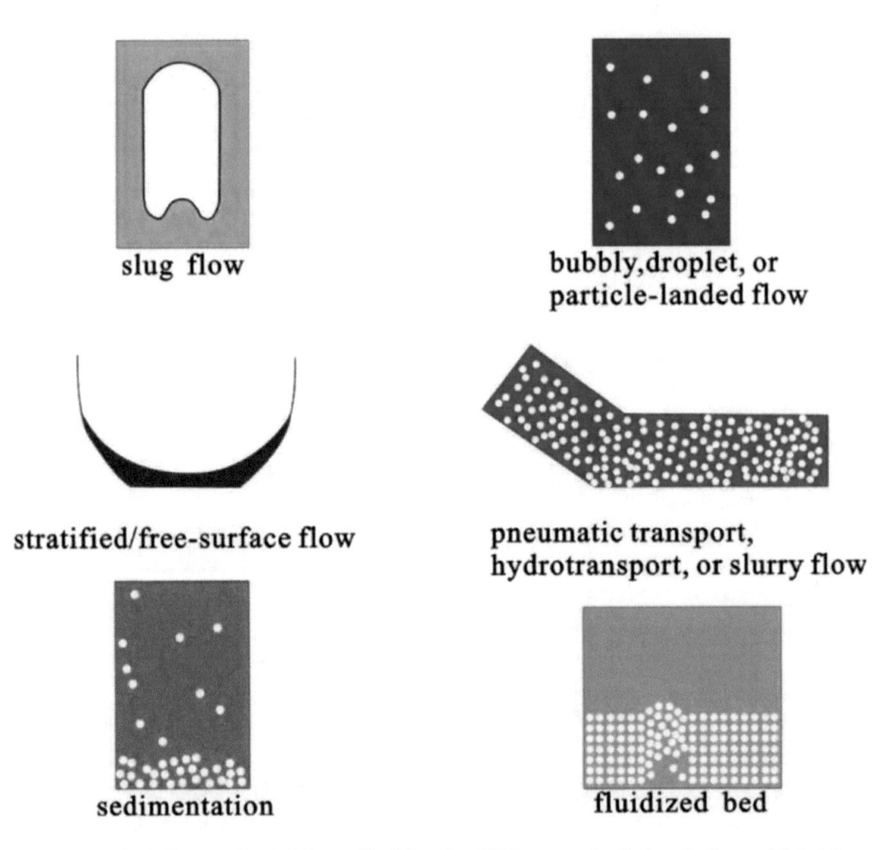

Figure 3. The Mixture model is applied for the CFD numerical simulation, which takes some relationships between two phases (fluid and solid particles) into account, including slug flow, bubbly, droplet, or particle-laden flow, stratified/free-surface flow, pneumatic transport hydrotransport, or slurry flow, sedimentation and fluidized bed.

Since the particulate loading of turbidity currents may vary from small to considerably large values, a Eulerian-Eulerian multiphase numerical approach is considered to be more appropriate, which can handle a wider range of particle volume fractions than a Eulaerian-Lagrangian approach (maximum particles volume fraction of 10-12%) (Cantero et al., 2008). FLUENT software provides various multiphase models that are based on the Eulerian-Eulerian approach. Considering that the VOF model is used to simulate the multiphase flow with free surface and the Euler model has a relatively high computational cost, this study applies Mixture model for the simulations. In this model, the relationship between two phases (fluid and solid particles) is taken into account, including slug flow, bubbly, droplet, or particle-laden flow, stratified/free-surface flow, pneumatic transport hydrotransport, or slurry flow, sedimentation and fluidized bed (Figure 3). In this multiphase model, the different phases are treated mathematically as interpenetrating continua and therefore the concept of phasic volume fraction is introduced, where the volume fraction of each phase is assumed to be a continuous function of space and time. The sum of the volume fractions of the various phases is equal to unity. Accordingly a modified set of momentum and continuity equations for each phase are solved. Pressure and inter-phase exchange coefficients are used in order to achieve the coupling for these equations. The coupling of granular (fluid-solid) flows is handled differently than in the case of non-granular (fluid-fluid) flows. For granular flows, the properties are obtained from application of the kinetic theory. The type of phases involved also defines the momentum exchange between the various phases.

The motion of the suspended sediment particles within a turbidity current as well as the motion generated in the ambient fluid are of a highly turbulent nature. In order to account for the effect of turbulence in the numerical simulations of the present investigation, the instantaneous governing equations are not applied directly but they are ensemble-averaged, converting turbulent fluctuations into Reynolds stress, which represents the effects of turbulence.

3.2. Governing Equations

The Mixture model is designed for two or more phases (fluid or particulate). As in the Eulerian model, the phases are treated as interpenetrating continua. The mixture model includes particle-laden flows with low loading, bubbly flows, sedimentation and cyclone separators. The mixture model can also be used without relative velocities for the dispersed phases to model homogeneous multiphase flow. The mixture model allows the phases to be interpenetrating. The volume fractions and for a control volume can therefore be equal to any value between 0 and 1, depending on the space occupied by phase q and phase p. Moreover, the mixture model allows the phases to move at different velocities, using the concept of slip velocities. The mixture model solves the continuity equation for the mixture, the momentum equation for the mixture, and the volume fraction equation for the secondary phases, as well as algebraic expressions for the relative velocities if the phases are moving at different velocities.

3.2.1. Continuity Equation

The continuity equation for the mixture is

$$\frac{\partial}{\partial t}(\rho_m)+\nabla\cdot(\rho_m\vec{\upsilon}_m)=\dot{m} \tag{1}$$

where is the mass-averaged velocity

$$\vec{\upsilon}_m=\frac{\sum_{k=1}^{n}\alpha_k\rho_k\vec{\upsilon}_k}{\rho_m} \tag{2}$$

and is the mixture density

$$\rho_m=\sum_{k=1}^{n}\alpha_k\rho_k \tag{3}$$

where is the volume fraction of phase k, and represents mass transfer due to cavitation or user-defined mass sources.

3.2.2. Momentum Equation

The momentum equation for the mixture can be obtained by summing the individual momentum equation for all phases. It can be expressed as

$$\frac{\partial}{\partial t}(\rho_m\vec{\upsilon}_m)+\nabla\cdot(\rho_m\vec{\upsilon}_m\vec{\upsilon}_m)=-\nabla_p+\nabla\cdot[\mu_m(\nabla\vec{\upsilon}_m+\nabla\vec{\upsilon}_m^T)]+\rho_m\vec{g}+\vec{F}+\nabla\cdot(\sum_{k=1}^{n}\alpha_k\rho_k\vec{\upsilon}_{dr,k}\vec{\upsilon}_{dr,k}) \tag{4}$$

where n is the number of phases, is a body force, and is the viscosity of the mixture:

$$\mu_m = \sum_{k=1}^{n} \alpha_k \mu_k \tag{5}$$

$\vec{\upsilon}_{dr,k}$ is the drift velocity for secondary phase k:

$$\vec{\upsilon}_{dr,k} = \vec{\upsilon}_k - \vec{\upsilon}_m \tag{6}$$

3.2.3. Relative (Slip) Velocity and Drift Velocity

The relative velocity (also referred to as the slip velocity) is defined as the velocity of a secondary phase p relative to the velocity of the primary phase q:

$$\vec{\upsilon}_{qp} = \vec{\upsilon}_p - \vec{\upsilon}_q \tag{7}$$

The drift velocity and the relative velocity ($\vec{v}_{qp}$) are connected by the following expression:

$$\vec{\upsilon}_{dr,p} = \vec{\upsilon}_{qp} - \sum_{k=1}^{n} \frac{\alpha_k \rho_k}{\rho_m} \vec{\upsilon}_{qk} \tag{8}$$

FLUENT's mixture model makes use of an algebraic slip formulation. The basic assumption of the algebraic slip mixture model is that, to prescribe an algebraic relation for the relative velocity, a local equilibrium between the phases should be reached over short spatial length scales. The form of the relative velocity is given by

$$\vec{\upsilon}_{qp} = T_{qp} \vec{\alpha} \tag{9}$$

where is the secondary-phase particle's acceleration and is the particulate relaxation time. Following Manninen et al. 1993, is of the form:

$$T_{qp} = \frac{(\rho_m \quad \rho_p) d_p^2}{18 \mu_q f_{drag}} \tag{10}$$

where is the diameter of the particles of secondary phase p, and the drag function is taken from Schiller and Naumann (1935):

$$f_{drag} = \begin{cases} 1+0.15\,\mathrm{Re}^{0.687} & \mathrm{Re} \leq 1000 \\ 0.0183\,\mathrm{Re} & \mathrm{Re} > 1000 \end{cases} \tag{11}$$

The acceleration is of the form

$$\alpha = g \quad (v_m \bullet \nabla) v_m \quad \frac{\partial v_m}{\partial t} \tag{12}$$

The simplest algebraic slip formulation is the so-called flux model, in which the acceleration of the particle is given by gravity and/or a centrifugal force and the particulate relaxation time is modified to take into account the presence of other particles.

3.2.5. Volume Fraction for the Secondary Phases

From the continuity equation for second phase p, the volume fraction equation for secondary phase p can be obtained:

$$\frac{\partial}{\partial t}(\alpha_p \rho_p) + \nabla \cdot (\alpha_p \rho_p \vec{\upsilon}_{dr,p}) = -\nabla \cdot (\alpha_p \rho_p \vec{\upsilon}_{dr,p}) \tag{13}$$

4. DESCRIPTION OF NUMERICAL SIMULATIONS

FLUENT is a Computational Fluid Dynamical (CFD) software package, which uses unstructured meshes in order to reduce the amount of time to generate meshes, simplify the geometry modeling and mesh generation process, and model more-complex geometries. FLUENT is capable of handling triangular and quadrilateral elements in 2D, and tetrahedral, hexahedral, pyramid, and wedge elements in 3D. This flexibility allows picking mesh topologies that are best suited for particular applications. First of all, we need to generate the initial mesh outside of the solver, using GAMBIT.

The basic procedural steps for numerical simulation via FLUENT can be outlined as creating the model geometry and grid, starting the appropriate solver for 2D or 3D modeling, importing the grid, checking the grid, selecting the solver formulation, choosing the basic equations to be solved, specifying material properties, specifying the boundary conditions, adjusting the solution control parameters, initializing the flow field, calculating a solution, examining the results, saving the results, and if necessary refining the grid or considering revisions of the numerical or physical model.

Hereby, we only introduce three key points for sedimentary processes of turbidity currents.

4.1. Create the Geological Model

Numerical simulation using FLUENT requires creating the geometry and meshing the models. First, we interpret the seismic data in QDNB and get the geometric parameters by back stripping strata, including slope gradient, slope length, slope heath and so on. According to the principles of geometric similarity, the geological model can be reduced in some ratio under the same internal Froude number (Fi) to create the numerical model in GAMBIT software. In this case, the status of fluid flow can keep the same. Then, we mesh the geometry in Gambit for CFD simulation.

4.2. Specify Boundary Conditions

Boundary conditions specify the flow variables on the boundaries of the numerical model. Therefore, they are a critical component of turbidity simulations and need to be specified appropriately. The boundary types available in FLUENT include flow inlet and exit boundaries, wall, repeating and pole boundaries, internal cell zones and internal face boundaries. According to the geological model, all necessary boundaries must be given properly.

4.3. Initialize the Flow Field

Before we start the calculations in meshed cells, we must initialize the flow field in the entire domain. We need to set initial values of velocities, fluid density, viscosity, grain size, grain density, rough bed, and so on. The outside environment, that is, the gravity field, also needs to be given. In this study, except for grain size, all other initial values are given according to published documents for observations of turbidity currents around the world and remain the same during the simulations. We only change the grain size to observe the geometries of turbidity deposits so that the component of submarine fans interpreted on seismic data can be predicted on the basis of the comparison of geometries between seismic data and forward simulation results.

5. Hydrodynamic Simulation for Reservoir Prediction in QDNB

5.1. Turbidity Current Generation

Integrated research for the distribution and sedimentary characteristics in QDNB show that the generations of turbidity currents are caused by relative sea level changes, tectonic activities and sediment supply, and the number and scale of deposits are related to the slope gradient (Zhang et al., 2012). There are a large amount of submarine fans composed of turbidites in the shelf-slope setting and few in the gentle slope setting. Moreover, the height of slope is the depth from the shoreline to the basin floor, which is usually deeper than 500 m, and the extensional distance in the geological model along the shoreline has almost no effect on the generation of turbidity currents. According to the principles of geometric similarity in hydrodynamic simulation, the fluid flow has the same status when reducing the scale with the same Froude number (Fr) as the one in the geological model. The Froude number (Fr) is defined as follows.

$$F_r = \frac{v}{\sqrt{gl}} \tag{14}$$

where v is the fluid velocity and l represents the slope length. Therefore, in the mathematic model, we only need to reduce the initial velocity and the mean square root of slope length in the same ratio, which can be obtained from the geological model interpreted from the seismic data. Assuming the depth from the shoreline to the basin floor is 500 m, we can get the slope gradient. In order to observe the mixture fluid flow in basin flow and the formation processes of submarine fans (which is the sedimentation of turbidity currents), we set a 50 m horizontal distance extending along the basin floor. Considering that the turbidity currents are usually triggered by the high-speed sedimentation of delta, we set the water depth of the inlet of fluid at 50 m. The specific scale of the numerical model is shown in Figure 4.

Under the same initial fluid density of turbidity currents, the forward simulations with various slope gradients and grain sizes illustrate that the generation of turbidity currents requires an appropriate slope gradient. If the slope is too gentle, the energy is not enough to suspend the particles and cannot cause turbidity currents. On the other hand, if the slope is too sharp, the solid particles deposit immediately and cannot result in large turbidity currents either. The appropriate slope gradient for the generation of turbidity current is about 1.5-3°, which coincides with the experiment conclusion of Pratson et al. (2000).

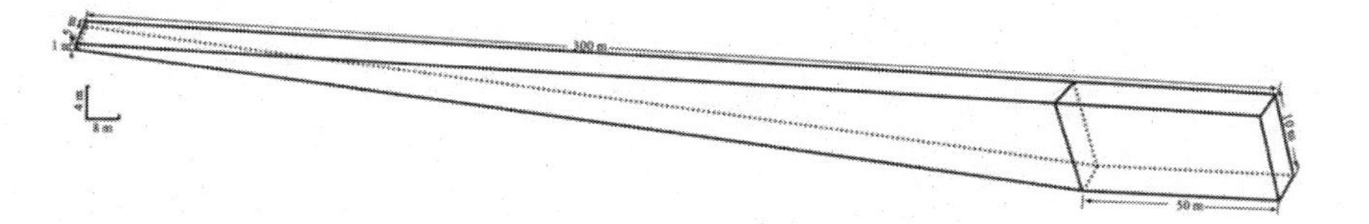

Figure 4. The Geological model applied in this turbidity simulation, which will be reduced to 1/50 for the mathematic model with the same Froude number. It indicates that the initial velocity of turbidity currents should be decreased according to the definition of Froude number.

5.2. Sedimentary Processes of Turbidity Current and Reservoir Prediction

The submarine fans interpreted from seismic data are usually comprised of gravity flow deposits, including turbidites, debrites, slump and so on. However only turbidites can behave as an excellent hydrocarbon reservoir. Under similar geological settings and provenances, the submarine fan could be the recurrence complex of one kind of gravity flow, therefore the grain size of one turbidity current can represent the component of the complex (submarine fan).

In the software FLUENT, the transported particles can be observed along their flow direction, which indicates that the location of sedimentation can be confirmed when the velocity decreases to zero (Figure 5). The simulations under various slope gradients, initial fluid density and grain sizes show that the turbidity current is generally formed as slope fans and/or basin floor fans, and the thickest segment is located in the slope root and its thickness is thinning toward the basin floor. Furthermore, under the same conditions, larger grain sizes generate thicker turbidites with a limited spread, and vice visa.

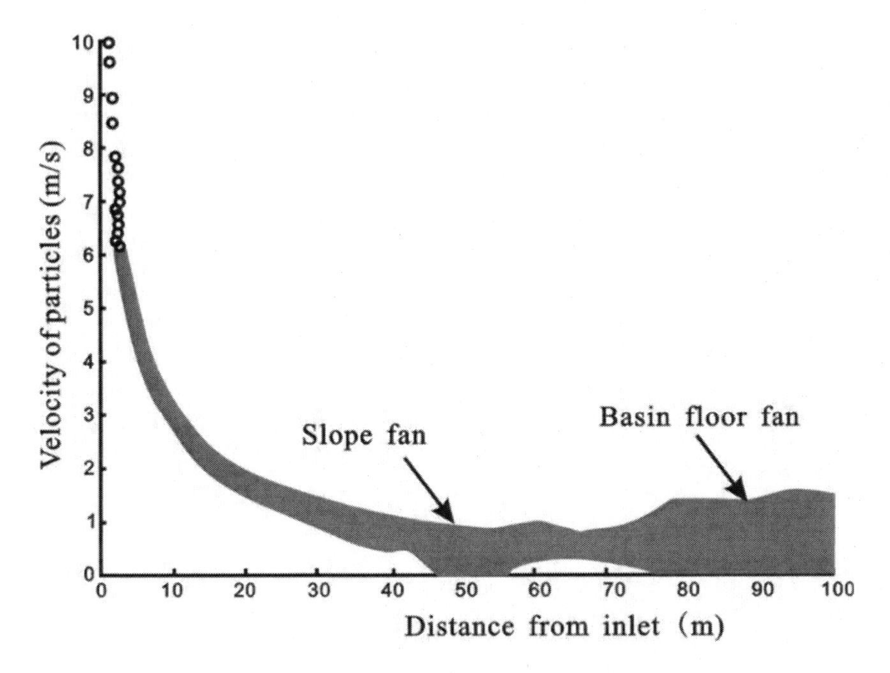

Figure 5. Velocity magnitude changes along turbidity flow. When it decreases to zero, the sediments will settle down. It means that the formation processes and sedimentary locations of the slope fan and the basin floor fan can be observed.

In order to verify the reservoir prediction method, this study takes two cases to simulate the sedimentary processes and predict their reservoir components in QDNB, which have been proved by drillings. The geological models are obtained by back stripping the strata interpreted from seismic data, and they are reduced 1/50 to get the mathematic model for simulation.

Given 20% particle content and 10 m/s initial velocity, the sedimentary processes of fluid flow and their geometries of deposits are simulated with grain sizes of 1 mm (sand), 0.1 mm (silt) and 0.01 mm (mud) respectively.

Figure 6 shows the simulation results in the case of 12.5 km slope length and 2.0° slope gradient. When the grain size is 1 mm (sand), it becomes lower stand wedge sedimentation (Figure 6A).

Actually, it starts to deposit since its source until all sediments are settled down with no turbulence, so it is not the turbidity current but debris. When the grain size is changed to 0.1 mm (silt), the turbulence can be observed and obvious slope fans and indistinct basin floor fans are formed (Figure 6B). When the grain size is 0.01 mm (mud), the typical slope fan and basin floor fan are deposited with violent turbulences (Figure 6C). Comparing these simulation results with the seismic refection features, the characteristics of a larger slope fan and a slight basin floor fan indicates that the turbidites should consist of silts, which coincide with the drilling result of Well A (Figure 7).

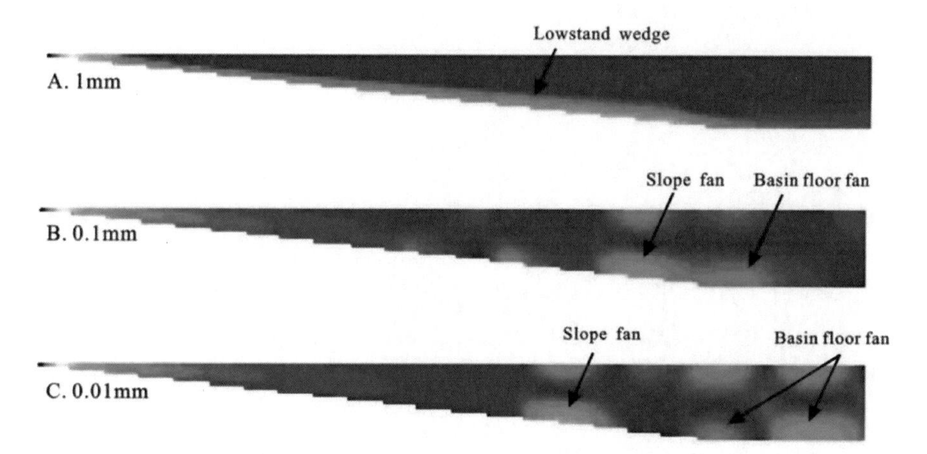

Figure 6. Geometric characteristics of deposits while various granularities are given. As described in the context, when the grain sizes are 0.01 mm, 0.1 mm and 1 mm respectively, the occurrence of turbulences and the formation of the slope and basin floor fan can be observed. The green color shows the low velocity and high particle content, which means sedimentation. And the blue color shows the high velocity and no sedimentation.

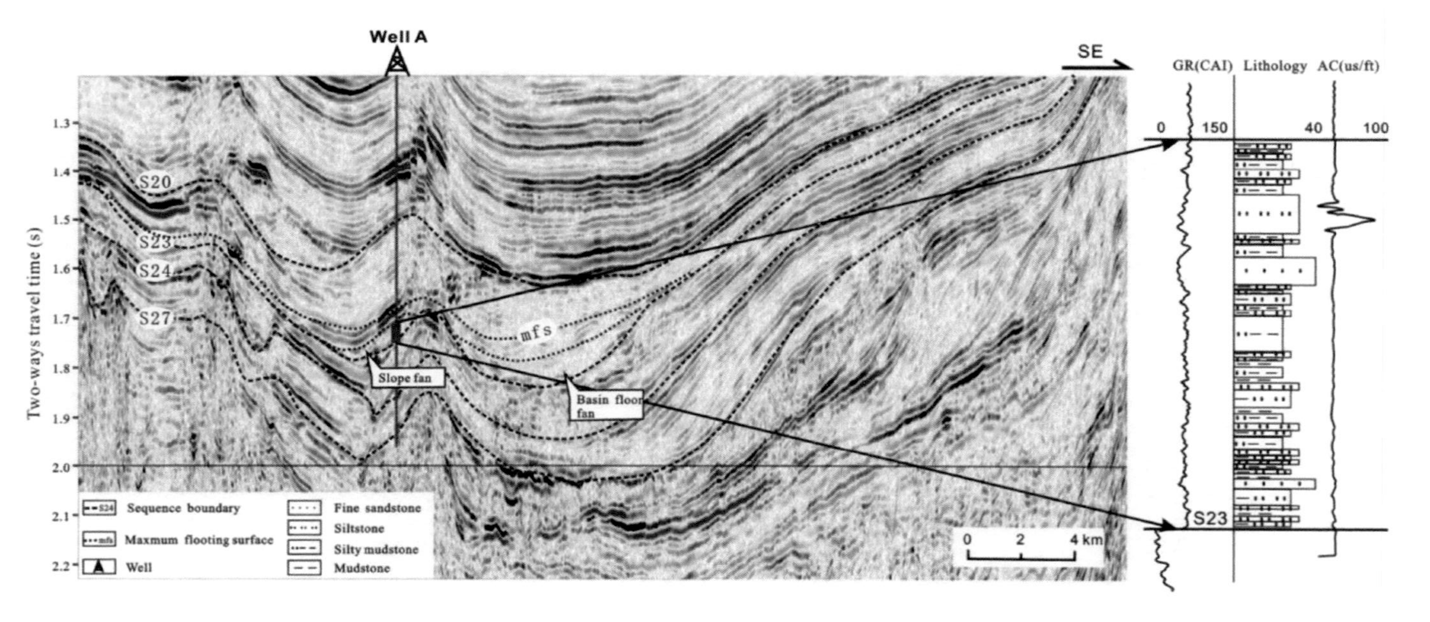

Figure 7. Turbidity sediments interpreted by the integration of seismic data and Well A in the Yinggehai Formation of QDNB. The locations of the seismic profile and well are seen in Figure 1. Compared with the simulation results, the component of the deposited fans should generally be siltstone, which is proved by Well A.

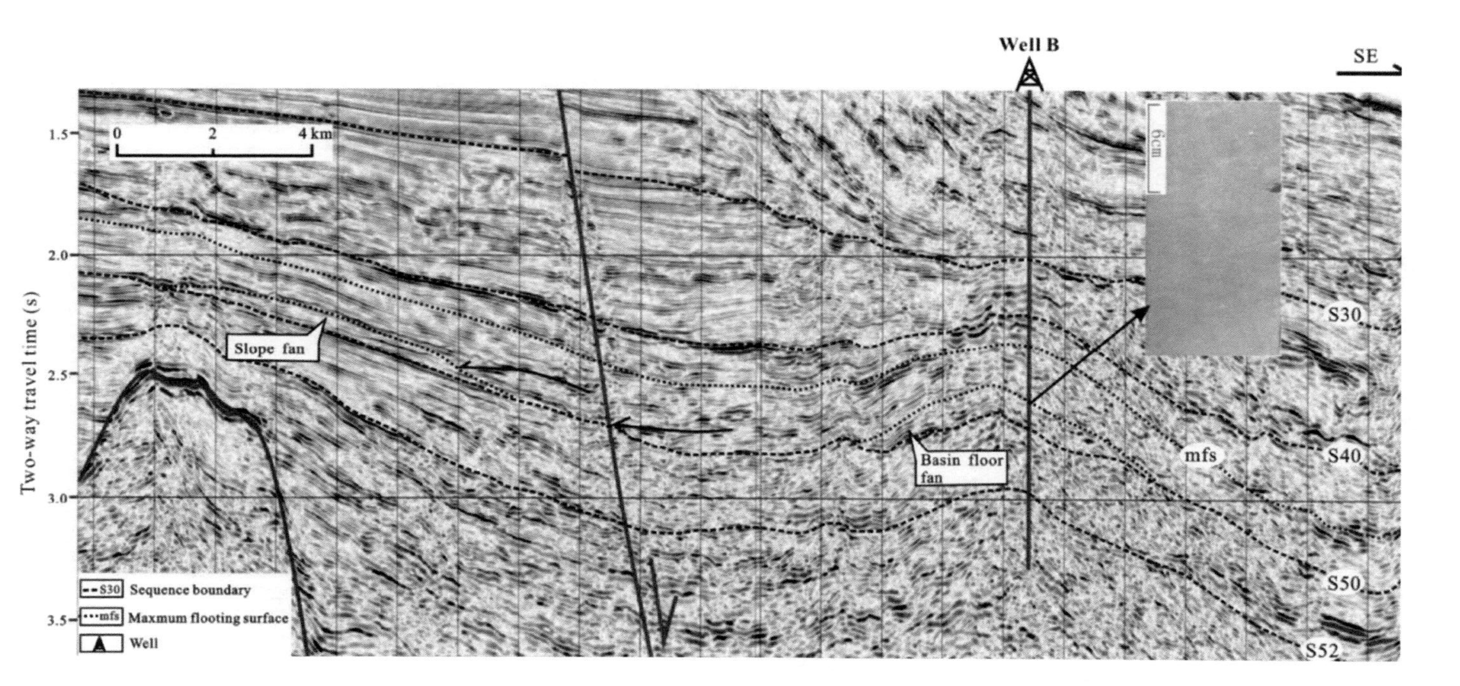

Figure 8. Turbidity sediments interpreted by the integration of seismic data and Well B in the Meishan Formation of QDNB. The locations of the seismic profile and well are seen in Figure 1. Compared with the simulation results shown in Figure 6, the basin floor fan should be comprised of mudstone and would not be a good reservoir, which coincides with the drilling result shown by the core image.

Another case with a 9 km slope length and a 2.9^0 slope gradient is also executed. Although its slope is sharper than the last one due to more tectonic subsidence and poor sediment supply, it is not enough to cause slumps and turbidity currents still develop, which have similar geometries to those in the last case under various grain sizes. According to the seismic interpretation, there are not only slope fans but also obvious basin floor fans. Based on the simulation results, the turbidites should be comprised of silts, which have been confirmed by the drilling result of Well B.

Discussion and Conclusion

The study proposes a new attempt to predict deepwater hydrocarbon reservoirs, and some drillings have confirmed its validity. However, the geological processes are complex, and the numerical simulation has simplified them greatly. Some of the ignored factors may have a great effect on the sedimentary processes of turbidity currents. For example, the fluid flow is assumed as a closed system without considering fluid and particles exchanges between fluid body and bed, around fluid by erosion and entrainment. Although some numerical simulations have taken them into account (Amos et al., 1992; Parchure et al., 1985; Parsons and Garcia, 1998; Bosse et al., 2006), it is still not fully understood since the erosion can be affected by complex natural processes such as bioturbation, storm reforming bed with sedimentary structure, grain size, density and consolidation, and so on. Moreover, the trigger mechanism and provenances may have slight changes during the formation of submarines fans even though they are under a similar geological setting, whereas the simulation assumes that they are the recurrences of the same sedimentary processes. Therefore, the simulation is only used to predict the main component and some other grain sizes may still exist in the whole sedimentary bodies. For example, in Well A, although it is mainly comprised of silts, there are intervals of fine sands. Therefore, when we take the numerical simulation into practical application, the multidisciplinary integration under a specific geological setting is necessary and important.

ACKNOWLEDGMENTS

The study is supported by the National Natural Science Foundation of China (No. 91028009 and 40806019) and by the Special Foundation for State Major Basic Research Program of China (No.2011ZX05025-0020-020-03). We would like to acknowledge the Zhanjiang Branch of China National Offshore Oil Corporation for providing geological data. And our thanks are given to Xiaofei Mao from Wuhan University of Technology for the license of the software FLUNET.

REFERENCES

Amos, C. L., Daborn, G. R., Christian, H. A., et al. (1992). In situ erosion measurements on fine-grained sediments from the Bay of Fundy. *Marine Geology*, *108*, 175-196.

Amy, L. A., Talling, P. J., Peakall, J., et al. (2005). Bed geometry used to test recognition criteria of turbidites and (sandy) debrites. *Sedimentary Geology*, *179*, 163-174.

Blanchette, F., Strauss, M., Meiburg, E., Kneller, B. & Glinsky, M. E. (2005). High-resolution numerical simulations of resuspending gravity currents: conditions for self-sustainment. *Journal of Geophysical Research*, *110*, C12022.

Bombardelli, F. A., Cantero, M. I., Buscaglia, G. C. & Gacoa, M. J. (2004). Comparative study of convergence of CFD commercial codes when simulating dense underflows. In: Buscaglia G., Dari E., Zamonsky O. (eds). Mec'anica Computacional, XXIII. Bariloche, Argentina.

Bosse, T., Kleiser, L. & Meiburg, E. (2006). Small particles in homogenous turbulence: *settling velocity enhancement by two-way coupling Physical fluids*, *18*, 027102.

Cantero, M. I., Balachandar, S. & Garcia, M. H. (2008). A Eulerian-Eulerian model for gravity currents driven by inertial particles. *International Journal of Multiphase Flow*, *34*, 484-501.

Chen, P. H., Chen, Z. Y. & Zhang, Q. M. (1993). Sequence stratigraphy and continental margin development of northwestern shelf of the South China Sea. *AAPG Bull*, *77*, 842-862.

Ericilla, G., Casas, D., Estrada, F., et al. (2008). Morpho sedimentary features and recent depositional architectural model of the Cantabrian continental margin. *Marine Geology*, *247*, 61-83.

Hao, S. S., Huang, Z. L., Liu, G. D. & Zheng, Y. L. (2000). Geophysical properties of cap rocks in Qiongdongnan Basin, South China Sea. *Marine and Petroleum Geology*, *17*, 547-555.

Hartel, C., Meiburg, E. & Necker, F. (2000). Analysis and direct numerical simulation of the flow at a gravity-current head: Part 1. Flow topology and front speed for slip and no-slip boundaries. *Journal of Fluid Mechanics*, *418*,189-212.

Heimsund, S., Hansen, E. W. M. & Nemec, W. (2002). Computation 3D fluid-dynamics model for sediment transport, erosion and deposition by turbidity currents. In: Knoper M., Cairncross B. (eds). Abstracts, international association of sedimentologists 16th international sedimentological congress, Rand Afrikaans University, Johannesburg, 152-152.

Imaran, J., Kassem, A. & Khan, S. M. (2004). Three-dimensional modelling of density current, I. Flow in straight confined and unconfined channels. *Journal of Hydraulogic Reseach*, *42*(6), 578-590.

Jiang, T., Xie, X. N., Tang, S. L., Zhang, C. & Du, X. B. (2007). Numerical simulation on the evolution of sediment waves caused by turbidity currents. *Chinese Science Bulletin*, *52*, 2429-2434.

Kassem, A. & Imran, J. (2001). Simulation of turbid underflows generated by the plunging of a river. *Geology*, *29*(7), 655-658.

Lavelli, A., Boillat, J. L. & De Cesare, G. (2002). Numerical 3D modelling of the vertical mass exchange induced by turbidity currents in Lake Lugano (Switzerland). In: Proceedings 5th international conference on hydro-science and –engineering (ICHE-2002).

Manninen, M., Taivassalo, V. & Kallio, S. (1996). On the mixture model for multiphase flow. VTT publications 288, Technical Research Centre of Finland.

Mehdizadeh, A., Firoozabadi, B. & Farhanieh, B. (2008). Numerical simulation of turbidity curren using turbulence model. *Journal of Application of Fluid Mechanics*, *1*(2), 45-55.

Meiburg, E. & Kneller, B. (2010). Turbidity Currents and Their Deposits. *Annual Review of Fluid Mechanics*, *42*, 135-156.

Necker, F., Hartel, C., Kleiser, L. & Meinburg, E. (2002). High-resolution simulations of particle driven gravity currents. *International Journal of Multiphase Flow*, *28*, 279-300.

Pang, X., Chen, C. M., Wu, M. S., He, M. & Wu, X. J. (2006). The Pearl River deep-water fan systems and significant geological events. *Advances in Earth Science*, *21*(8), 793-799 (in Chinese with English abstract).

Parchure, T. M. & Mehta, A. J. (1985). Erosion of soft cohesive sediment deposits. *Journal of Hydraulic Engineering*, *111*, 1308-1326.

Parsons, J. D. & Garcia, M. (1998). Similarity of gravity current fronts. *Physical fluids*, *10*, 3209-3213.

Pettingill, H. S. & Weimer, P. (2002). World wide deep water exploration and production: past, present, and future. *The Leading Edge*, *21* (4), 371-376.

Posamentier, H. W. & Kolla, V. (2003). Seismic geomorphology and stratigraphy of depositional elements in deep-water settings. *Journal of Sedimentray Research*, *73*, 367-388.

Pratson, L. F., Imran, J., Parker, G., et al. (2000). Debris flows versus turbidity currents: a modeling comparison of their dynamics and deposits. In: Bouma A H, Stone C G. Fine-grained turbidite systems. *AAPG Memoir 72 /SEPM Special Publication*, *68*, 57-72.

Schiller, L. & Naumann. (1935). *Z. Ver. Deutsch. Ing*., *77*, 318.

Singh, J. (2008). Simulation of suspension gravity currents with different initial aspect ratio and layout of turbidity fence. *Application of Mathematic Model*, *32*, 2329-2346.

Su, M., Xie, X. N., Li, J. L., Jiang, T., et al. (2011). Gravity flow on slope and abyssal systems in the Qiongdongnan basin, northern South China Sea. *Acta Geologica Sinica* (English Edition), *85*(1), 243-253.

Wu, S. G., Yuan, S. Q., Zhang, G. C., Ma, Y.B., Mi, L. J. & Xu, N. (2008). Seismic characteristics of a reef carbonate reservoir and implications for hydrocarbon exploration in deepwater of the Qiongdongnan basin, northern South China Sea. *Marine and Petroleum Geology*, *26*(6), 817-823.

Xie, X. N., Müller, R. D., Li, S. T., Gong, Z. S. & Steinberger, B. (2006). Origin of anomalous subsidence along the northern South China Sea margin and its relationship to dynamic topography. *Marine and Petroleum Geology*, *23*(7), 745-765.

Xie, X., Müller, R. D., Ren, J., Jiang, T. & Zhang, C. (2008). Stratigraphic architecture and evolution of the continental slope system in offshore Hainan, northern South China Sea. *Marine Geology*, *247*, 129-144.

Yuan, S. Q., Yao, G. S., Lv, F. L., et al. (2009). Features of Late Cenozoic deepwater sedimentation in southern Qiongdongnan basin, northwestern South China Sea. *Journal of Earth Science*, *20*(1), 172-179.

Zhang, N., Jiang ,T. & Zhang, D. J. (2012). Topography and its control over deepwater sedimentation in the Qiongdongnan basin. *Marine geology and quaternary geology*, *32*(5), 27-33. (in Chinese with English abstract)

In: Basins
Editor: Jianwen Yang

ISBN: 978-1-63117-510-7

Chapter 3

SIMILARITIES AND DIFFERENCES BETWEEN 2-D AND 3-D NUMERICAL RESULTS OF ORE-FORMING FLUID FLOW IN SEDIMENTARY BASINS: EXAMPLE FROM MOUNT ISA BASIN, NORTHERN AUSTRALIA

Jianwen Yang[1,2,*], Baocheng Pang[1] and Zuohai Feng[1]

[1]College of Earth Sciences, Guilin University of Technology, Guilin, Guangxi, China

[2]Department of Earth and Environmental Sciences, University of Windsor, Windsor, Ontario, Canada

ABSTRACT

This chapter aims at addressing the major similarities and differences between 2-D and 3-D numerical modeling results of ore-forming hydrothermal fluid flow in sedimentary basins, with Mount Isa Basin, northern Australia as an example. Our numerical results of Sedex-type deposits indicate that both the 2-D and 3-D fluid flow and hydrothermal discharge are controlled by the spatial relationships between active

* Corresponding author: Tel: (519)253-3000/ext 2181; Fax: (519)973-7081; Email: jianweny@uwindsor.ca.

synsedimentary faults and clastic aquifers. However, the general recharge-discharge pattern of fluid flow established on the basis of 2-D modeling is usually not valid under 3-D conditions, even for a quasi 3-D domain constructed simply by extrapolating the 2-D cross-section along the longitudinal direction. This is because fluid tends to circulate within more permeable fault zones and form a series of planar convection cells over the fault planes rather than in less permeable host rocks, unless the faults and host rocks have a similar permeability range. Including a secondary cross fault to the 3-D model makes the fluid flow pattern as well as the heat regime become more distinct from the 2-D modeling result. Localized, subvertical columns of enhanced permeability, related to the intersection of the primary major faults with the secondary cross fault, are essential to developing 3-D 'mushroom-shaped' hydrothermal convection rolls that allow significant amount of fluids to circulate through the host rocks and leach sufficient metal content to form deposits of this type. Salinity variation through a basin has important implications for hydrothermal fluid flow, either promoting or impeding buoyancy-driven free convection. Our modeling results indicate that Sedex-type deposits are more easily formed when evaporation first produces surface brines and then these brines sink and displace pore waters in basins.

Keywords: Hydrothermal fluid flow, Ore-forming processes, Sedimentary basins, Fluid flow modeling

1. INTRODUCTION

A significant proportion of Earth's metallic mineral resources are hosted by sedimentary basins, which are estimated to contain about 75% of reserves (Cuney and Kyser, 2008). The development and refinement of genetic and exploration models for ore deposits in sedimentary basins require an understanding of hydrothermal fluid flow in this geological environment. Numerical modeling has proven to be an efficient tool for testing, comparing and contrasting various hypotheses regarding hydrothermal fluid flow and the resulting ore-forming processes. Computer models simulating fluid flow, heat transfer and mass transport in complex hydrothermal systems can provide considerable insight into how these systems operate to produce economic concentration of metals (e.g., Cui et al., 2012a; Beiraghdar and Yang, 2014).

Previous numerical studies usually considered 2-D models simulating selected geological sections (e.g., Bethke, 1986; Hobbs et al., 2000; Yang et al., 2006; Zhang et al., 2006; Zhang et al., 2010; Ju and Yang, 2011, Cui et al.,

2012b), which has greatly improved our understanding of ore-forming hydrothermal fluid flow in sedimentary basins. However, 2-D numerical simulation cannot fully encompass the complexities of 3-D hydrothermal system in reality. Recent fast developments in computer hardware and software have enabled researchers to start with simulating ore-forming hydrothermal fluid flow in realistic 3-D environments. For instance, Yang (2006) presented a 3-D hydrological model that fully couples transient fluid flow and heat transport in the McArthur basin, northern Australia, but without considering the effect of salinity distribution on regional fluid flow. Schaubs et al. (2006) developed a coupled 3-D deformation-fluid flow numerical model and investigated gold mineralization around basalt domes in the Stawell corridor. Zhao et al. (2008) considered the morphological evolution of 3-D chemical dissolution fronts occurring in fluid-saturated porous media. Feltrin et al. (2009) simulated 3-D fluid flow driven by tectonic deformation associated with the formation of the giant lead-zinc-silver Century deposit, but no account was taken of the effect of temperature and salinity variation on fluid flow in the hydrothermal system. More recently Yang et al. (2010) conducted a 3-D numerical investigation that fully couples transient fluid flow and heat transfer with solute transport pertinent to the genesis of shale-hosted lead-zinc ores in the Mount Isa basin, northern Australia.

The principal aim of this chapter is to present a number of unpublished 2-D and 3-D modeling results of hydrothermal fluid flow associated with the formation of Sedex-type deposits in the Mount Isa basin, northern Australia as a complement to our previous studies, and in particular to address the major similarities and differences between the 2-D and 3-D numerical simulations and to highlight the effect of salinity variations on ore-forming hydrothermal fluid flow under 3-D conditions.

2. Governing Equations and Finite Element Modeling

Buoyancy-driven fluid flow is recently gaining momentum as the most likely hydrological scenario for the formation of sedimentary-exhalative (SEDEX) ore deposits (e.g., Yang et al., 2006; Yang et al., 2009). Mathematical equations that govern the physical behavior of this type of fluid flow in sedimentary basins can be derived by considering the continuity of fluid flow, heat energy and solute transport, in conjunction with Darcy's law.

Detailed derivation of these equations can be found in standard textbooks (e.g., Bear, 1972). The fluid continuity equation can be expressed in term of an 'equivalent freshwater' hydraulic head for the variable density fluid flow system (Frind, 1982). For a 3-D system in the x-y-z coordinate, we have:

$$\frac{\partial}{\partial x}\left(K\frac{\partial h}{\partial x}\right)+\frac{\partial}{\partial y}\left(K\frac{\partial h}{\partial y}\right)+\frac{\partial}{\partial z}\left(K\frac{\partial h}{\partial z}+K\rho_r\right)=S_s\frac{\partial h}{\partial t}, \tag{1}$$

$$\frac{\partial}{\partial x}(\lambda_m\frac{\partial T}{\partial x})+\frac{\partial}{\partial y}(\lambda_m\frac{\partial T}{\partial y})+\frac{\partial}{\partial z}(\lambda_m\frac{\partial T}{\partial z})$$
$$-\frac{\partial}{\partial x}(c_w\rho_w q_x T)-\frac{\partial}{\partial y}(c_w\rho_w q_x T)-\frac{\partial}{\partial z}(c_w\rho_w q_z T)=c_m\rho_m\frac{\partial T}{\partial t}, \tag{2}$$

and

$$\frac{\partial}{\partial x}(\theta D_x\frac{\partial C}{\partial x})+\frac{\partial}{\partial y}(\theta D_y\frac{\partial C}{\partial y})+\frac{\partial}{\partial z}(\theta D_z\frac{\partial C}{\partial z})$$
$$-\frac{\partial}{\partial x}(q_x C)-\frac{\partial}{\partial y}(q_y C)-\frac{\partial}{\partial z}(q_z C)=\frac{\partial}{\partial t}(\theta RC), \tag{3}$$

where q_x, q_y and q_z are the Darcy flux components in *x, y,* and *z* directions, *h* the 'equivalent freshwater' head, *K* the hydraulic conductivity, S_s the specific storage, $\lambda_m = \lambda_w{}^{\theta}\lambda_s{}^{(1-\theta)}$ (λ_w and λ_s denote the thermal conductivity of the fluid and solid phase), θ the porosity, $c_m\rho_m = c_w\rho_w\theta + c_s\rho_s(1-\theta)$ (c_w and c_s are the specific heat capacity of the fluid and solid phase, and ρ_w and ρ_s the density of the fluid and solid phase), ρ_r the relative fluid density defined as $\rho_r = \rho_w/\rho_0 - 1$ (ρ_0 is the reference fresh water density), *t* the time, *C* the solute concentration, *T* the temperature, *R* the retardation factor, and D_x, D_y and D_z are the hydrodynamic dispersion coefficients. The governing equations (1) to (3) are not alone sufficient. The dependence of fluid density and viscosity on temperature and solute concentration must be defined. In this chapter it is calculated using the NIST/ASME Steam Properties code (Klein and Harvey, 1996).

These equations form a time-dependent, nonlinear and coupled system, which leads their solutions to become nontrivial even for a very simple 1-D geological system. We have recently developed a finite element computer package to numerically solve these governing equations, employing the standard Galerkin finite element technique combined with the symmetric matrix time integration scheme of Leisman and Frind (1989). The Leisman and Frind scheme is particularly useful as it generates a symmetric coefficient matrix for both fluid flow and the heat/mass transport equations and hence reduces array storage requirements. We adopt a non-orthogonal quadrilateral mesh to create finite elements since it is better suited for the complex geometry of different stratigraphic units and faults encountered in sedimentary basins. Simulation domain is discretized to create a series of finite elements (or cells), and each element is assigned permeability, porosity, thermal conductivity and other physical properties governing hydrothermal fluid flow. The computer package is capable of modeling geologically reasonable hydrothermal systems containing uneven surface topography, irregularly-shaped rock units and freely-oriented fracture and fault zones. The details of the finite element algorithm and its application can be found in our previous publications (e.g., Yang et al., 1998; Yang, 2011).

3. Conceptualized Model and Modeling Results

Figure 1 illustrates the paleo-hydrological model over a central x-z cross-section based on the Mt. Isa basin, which has a dimension of 60 km × 20 km and is discretised by a 2-D non-orthogonal quadrilateral mesh consisting of 52 element columns and 19 rows. The conceptual model is constrained by some of the common features of a sedimentary basin's rift-and-sag phase, and in particular by the reconstructions of the Mount Isa basin, northern Australia (O'dea et al., 1997; Betts et al., 2003). It represents the highly conceptualized and simplified subsurface stratigraphy and structure that controlled the hydrological system when the Mount Isa Sedex deposits were formed early in the history of the Mount Isa basin.

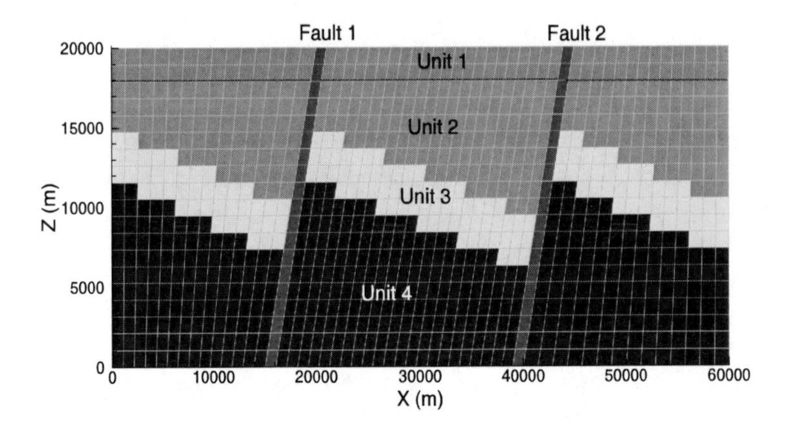

Figure 1. Highly simplified model of titled and submerged rift blocks associated with the formation of Sedex deposits in the Mount Isa basin, northern Australia, superimposed over the finite element mesh. See Table 1 for the hydrological and thermal properties assigned to the faults and host rock units.

Table 1. Major physical parameters assigned for the conceptualized 2-D and 3-D models

Hydrological unit and formation	Horizontal Permeability (m^2)	Vertical Permeability (m^2)	Thermal Conductivity ($J\ m^{-1}\ s^{-1}\ {}^{o}C^{-1}$)
Fault	4×10^{-14}	4×10^{-14}	2.0
Unit 1	1×10^{-15}	1×10^{-17}	2.0
Unit 2	2×10^{-15}	2×10^{-17}	2.5
Unit 3	4×10^{-14}	4×10^{-16}	3.0
Unit 4	2×10^{-16}	2×10^{-18}	3.0

The model involves a volcanic basement sequence of low permeability (Unit 4), a sandstone aquifer of high permeability (Unit 3), a rift cover sequence of intermediate permeability (Unit 2), and an upper cover sequence of shales and siltstones (Unit 1) that hosts mineral deposits formed during or soon after sediment deposition. Two permeable faults (Fault 1 and Fault 2) are also included, which penetrate from the upper sequence into the basement. The faults are 1 km wide and steeply dipping, characteristics that are constrained by surface exposure and seismic profiling (Bierlein and Betts, 2004). As with the previous studies of McLellan et al. (2006) and Oliver et al. (2006), Fault 1 on the left is equivalent to the Mount Isa fault system, and Fault 2 on the right simulates a fault zone farther north, such as the Termite Range fault at the Century deposit. The two steeply dipping faults cut the sandstone aquifer (Unit 2), forming a favorable hydrological framework for regional-scale fluid flow. The permeabilities and thermal conductivities assigned to the faults and host rocks are given in Table 1, following the previous numerical investigations in the Mount Isa-McArthur basin region (e.g., Yang et al., 2004a,b; Yang, 2006; McLellan et al., 2006; Oliver et al., 2006). We assume that the vertical permeability of the host rocks is two orders of magnitude less than the horizontal permeability due to the stratified nature of the sedimentary rocks. In addition, we assume $c_w = 4174$ J / kg°C, $c_s = 800$ J / kg°C, $\lambda_w = 0.5$ W / m°C, $\rho_s = 2630$ kg / m^3, $\rho_0 = 1000$ kg / m^3, and ϕ=10 %. The upper boundary (i.e., the basin floor) is permeable to fluid flow and fixed at 20°C. Over the top of the faults, the vertical temperature gradient is fixed at zero. That is, the fluid is assumed to be isothermal near the top surface owing to fluid flow via the faults. The lower boundary is maintained at 450°C, justified by present-day heat flow measurements in northern Australia (S. McLaren and M. Sandiford, pers. comm., 2001). The bottom of the model is assumed impermeable since it lies within the deep volcanic basement.

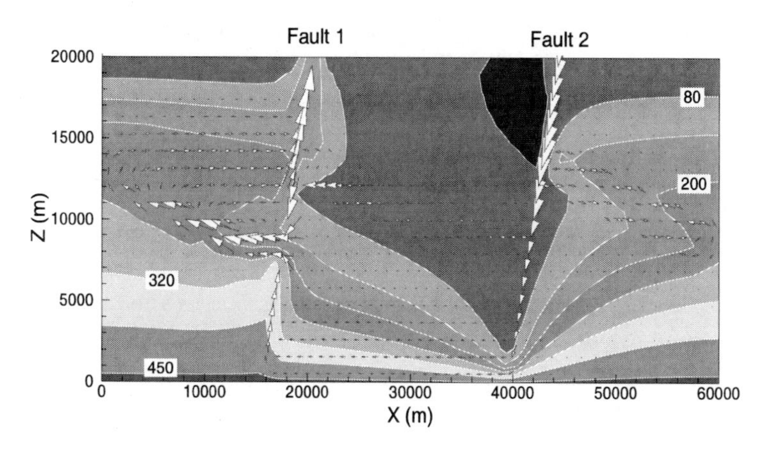

Figure 2. 2-D numerical modeling results of temperature distribution and fluid velocity vectors at 0.8Ma.

The side boundaries are assumed to be adiabatic to heat transfer and impermeable to fluid flow. The initial fluid velocity is set to zero over the whole solution domain, and the initial temperature is assumed to vary linearly with depth.

Figure 2 shows the temperature and fluid velocity distribution at 0.8 Ma, indicating that the cold seawater penetrates downwards along Fault 2 and then flows laterally, mainly through the permeable aquifer but with some flow via the less permeable basement sequence at depth. As the system develops, the fluids are heated from below. Fluids with elevated temperature ascend along Fault 1 driven by the buoyancy force and ultimately discharge onto the seafloor where they could form a Sedex-type deposit, such as the Mount Isa deposit that occurs adjacent to the Mount Isa fault. The model venting fluid temperature and velocity are 160°C and 4.1 m/year, respectively. However, this 2-D numerical modeling does not take into account longitudinal fluid flow and thermal transport along the strike direction of the faults, and essentially 'enforces' fluids to circulate over the transverse cross-section with fluids penetrating downwards along the recharge fault (Fault 2) from the basin floor, traveling laterally through the aquifer and surrounding rocks, and finally discharging back to the basin floor via the discharge fault (Fault 1). This dominant recharge-discharge pattern may not exist under more realistic 3-D conditions.

Let us develop a simple, quasi 3-D model by extrapolating the 2-D cross section on the x-z plane (Figure 1) for 30 km along the y-direction and assume that the structural and sedimentary facies architecture, as well as their physical properties, remains unchanged over the longitudinal dimension. The model is now discretized by a 3-D non-orthogonal quadrilateral mesh consisting of 21,736 finite elements. Figure 3 illustrates the fluid velocity vectors and temperature distribution at 0.8 Ma. It can be seen that due to the large contrast of permeabilities between the faults and host rocks, significant fluid circulation is confined mainly within the more permeable faults, with a maximum rate of 2.3 m/year, rather than in the host media (Figure 3a). Both upwelling and downwelling flow develops within these two faults at different longitudinal distances, which gives rise to significant temperature variation along the fault strike direction (Figure 3b). By comparison, fluid flow in the host rocks, including the aquifer, is less significant than in the 2-D case, as is evidenced by the nearly flat isotherms (Figure 3b). Since hydrothermal fluid flow more easily focuses and circulates within the fault zones with less fluid transport in the host rocks, the metal contents leached from the host rocks by the circulating basinal brines are likely insufficient to form giant Sedex deposits. This situation may change if the sandstone aquifer (Unit 2) becomes more permeable. Our previous numerical experiments (Yang et al., 2010) indicate that when the permeability contrast between the faults and aquifer is less than one order of magnitude, a 3-D convective fluid flow system can develop over a large volume of the host rocks, allowing for leaching of significant quantities of metals from the source rocks, which consequently is favourable for the formation of Sedex-type deposits.

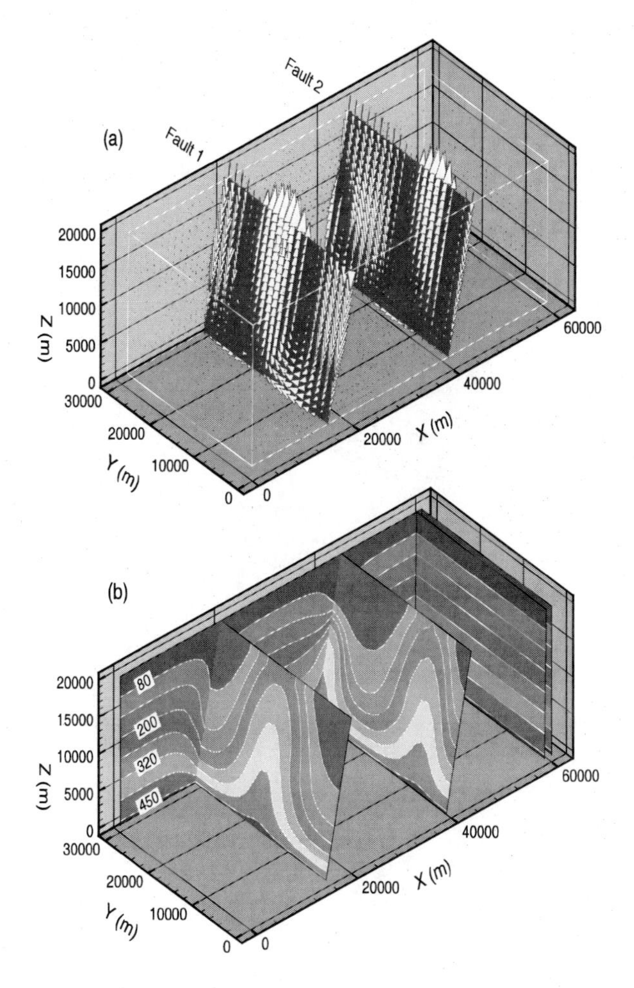

Figure 3. 3-D numerical modeling results at 0.8Ma: (a) fluid velocity vectors; and (b) temperature distribution. Note that the fluid velocity in the host rocks is too small to be visualized.

However, when the permeability contrast exceeds one order of magnitude, fluid flow tends to be confined within the more permeable fault zones as 2-D convection cells, as shown in Figure 3a.

In order to assess the overall effect of several minor faults that sinuously and intermittently cut the longitudinal major faults in the targeted region, let us now include a vertical transverse fault and keep other conditions the same as in the quasi 3-D model.

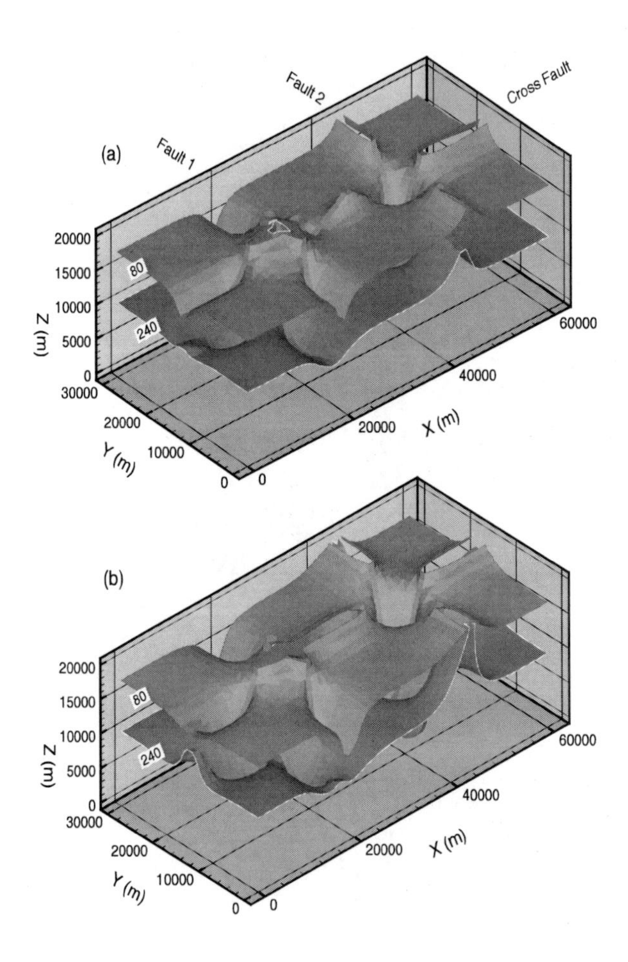

Figure 4. 3-D numerical modeling results of iso-surface of temperature distribution, corresponding to the condition that a cross fault is included to intersect the two major faults: (a) at 0.5 Ma; and (b) at 0.8 Ma.

The cross fault, located in the central x-z plane at y=15 km, orthogonally intersects the two major faults. To simplify the situation, this cross fault is assumed to have the same width and hydraulic conductivity as those of the primary main faults. The intersection of these two sets of faults may result in localized zones of enhanced permeability; therefore, two subvertical columns, centered on the left- and right-hand intersections and having cross-sectional areas of 3 km by 3 km, are included and assigned the same permeability as the faults. Figure 4 shows the 80°C and 240°C isotherms at 0.5 Ma and 0.8 Ma. It can be seen that fluids are mainly channelled downwards via the right-hand column and upwards through the left-hand column. This leads to the development of two major 3-D 'mushroom-shaped' convection rolls that are centered on the two columns of enhanced permeability. Significant hydrothermal convection now takes place, not only along the longitudinal direction as shown in Figure 3, but also along the transverse direction through both the cross fault and the host rocks, as evidenced by several upwelling and downwelling thermal plumes distributed alternately over the cross fault. The Fault 2-associated intersection column behaves as the major recharge pathway, whereas the Fault 1-related column serves as the major discharge conduit. Cold seawater moves downwards along the right subvertical column, then travels laterally to the left through the transverse fault and the aquifer (the host rock Unit 2). As the fluids flow through the system, they are heated from below, and eventually the hot brines ascend via the left subvertical column and discharge to the basin floor to form a Sedex-type deposit. The venting fluid temperatures on the basin floor range from 220°C to 240°C with fluid velocities of 1.9 to 3.5 m/year over the period of 0.8 Ma.

In these numerical experiments, the fluids were assumed to be pure water, with no account taken of the effect of salinity variation on fluid flow. High salinity (evaporitic) conditions have long been recognized as an important factor in the development of the brines that form stratiform lead-zinc ore deposits, mainly because the solubility of the economically important metals increases with increasing salinity (e.g., Hanor, 1996). We therefore next consider a semi-evaporitic condition, presumably resulting from the mixing of minimally modified seawater with saline brines generated by evaporation, by doubling the salinity of typical seawater 3.57% for the basin floor and assigning the seawater salinity for the lower boundary. Initial salinity is also equal to 3.57% throughout the basin, and other conditions remain the same as those in the quasi 3-D model. Figure 5 illustrates the salinity and temperature distribution at 0.25 Ma. It can be seen that the incorporation of saline fluids facilitates buoyancy-driven free convection, compared with the pure water case shown in Figure 3. Due to the continuous supply of brine from the basin floor, and hence the enhanced convection strength, saline fluid circulates not only within the faults but also spreads out via the aquifer and host rocks. Consequently, several 3-D convection rolls develop, which allows the hot brines to circulate through sufficient volumes of host rocks, and would facilitate the leaching of more metal and the development of deposits. Also, as illustrated in Figure 5b, several elongated discharge zones develop on the top of the faults, over which Sedex-type deposits may form.

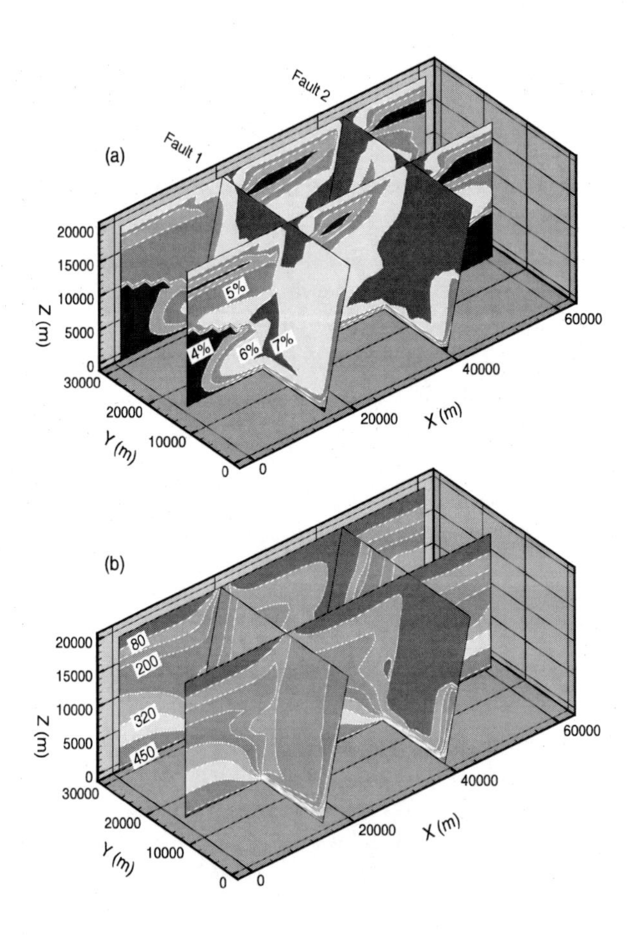

Figure 5. 3-D numerical modeling results at 0.25 Ma, corresponding to the semi-evaporitic condition: (a) salinity distribution; and (b) temperature distribution.

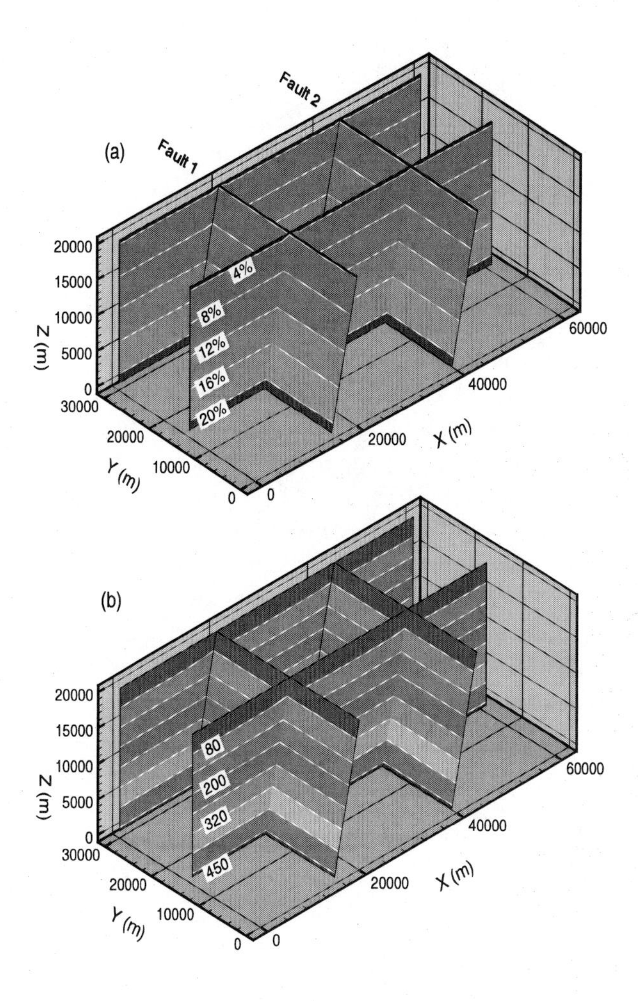

Figure 6. 3-D numerical modeling results at steady state, corresponding to the US Gulf Coast-type salinity condition: (a) salinity distribution; and (b) temperature distribution.

The venting fluid temperatures on the basin floor range from 180°C to 210°C with fluid velocities of 47 to 75 m/year over the period of 0.25 Ma. If the salinities of the formation waters in the basin increase with depth, from 3.57% at the basin floor to 20% at the bottom of the sequence, as is the case in the US Gulf Coast basin (e.g., Hanor, 1996), convective flow is considerably reduced and no discharge occurs along the basin faults (Figure 6), a scenario that is unfavorable for the formation of Sedex-type deposits. In order to initiate and maintain convective fluid circulation in this situation, a higher geothermal gradient is required such that the thermally induced buoyancy force can overcome the salinity-induced density force.

The numerical results presented here reveal that the spatial relationships between active synsedimentary faults and clastic aquifers control basinal fluid flow and hydrothermal discharge in both the 2-D and 3-D environments. However, the general fluid flow pattern, and hence the thermal regime, obtained using 2-D simulations may not be valid under 3-D conditions. This is because significant fluid circulation takes place within the more permeable faults in the 3-D case, rather than in the aquifer and other host rocks, unless the permeability contrast between the faults and the aquifer is less than one order of magnitude. Our numerical experiments also indicate that increased salinity on the basin floor, resulting from evaporation, facilitates buoyancy-driven free convection and can expel pore waters through discharge zone(s) to form Sedex-type deposits at basin floor by displacing lower salinity pore waters in the basin. If only the longitudinal major faults are present, 3-D hydrothermal convection rolls tend to be stretched along the fault strike direction so as to form elongated discharge zones on the basin floor. Including a transverse fault to intersect the primary major faults leads to the development of more rounded 3-D convection rolls that allows the hot basinal brines to circulate through larger volumes of the host rocks to leach enough metal content, and more importantly it enables fluids to channel downwards to a greater depth and eventually discharge to the basin floor via a more 'localized' discharge zone with higher temperature and venting velocity, which is favourable for the formation of giant Sedex deposits. Our results also show that if the basinal fluids (formation waters) resident on the sedimentary column at depth are highly saline, a relatively high thermal gradient will be required in order for thermally-induced buoyancy to overcome the salinity-induced density effects. The geological implication of these numerical modeling results is that Sedex-type deposits are more easily formed when evaporation produces surface brines, which then sink and displace pore waters in the basin. Also, determining the spatial distribution of fault intersection zones may be critical to exploring for deposits of this type in sedimentary basins.

CONCLUSION

The general geological structure of the Mount Isa basin in northern Australia has been used to constrain our 2-D and 3-D conceptualized models, involving a sandstone aquifer and three other stratigraphical sequences cut by two steeply-dipping longitudinal fault zones and a secondary transverse fault. Our numerical simulations have revealed that the interplay between the active synsedimentary faults and clastic aquifer unit controls the basinal fluid flow and hydrothermal discharge in both the 2-D and 3-D numerical case studies. However, the fluid flow pattern and heat regime established on the basis of 2-D modeling are usually not valid under more realistic 3-D conditions. Therefore more cautions are required when making decisions based on 2-D numerical simulations.

Our numerical studies have demonstrated that fluid flow modeling is capable of producing considerable insight into how complex hydrothermal systems operate to produce economic concentration of metals under different geological conditions, which can help asses various ore genesis models. Our numerical experiments have also confirmed that fluid flow modeling can identify favorable and/or unfavorable hydrological, thermal and structural conditions for ore genesis, which can guide field exploration to target new ore bodies. However, it should be pointed out that the applicability of numerical modeling as a predictive tool for targeting prospective deposits can be achieved only for regions where systematical geological, geochemical, geophysical and structural datasets are available to constrain fluid flow models.

ACKNOWLEDGMENTS

This work was sponsored by the *Program to Sponsor Teams for Innovation in the Construction of Talent Highlands in Guangxi Institutions of Higher Learning*. Research was also partially supported by the Natural Sciences and Engineering Research Council of Canada (NSERC) through a Discovery Grant (RGPIN 261283) and by the Guangxi Natural Science Foundation through a research grant (2012GXNSFAA053180) to J. Yang.

REFERENCES

Bear, J., 1972. Dynamics of Fluids in Porous Media. American Elsevier, New York.

Beiraghdar, Y., and Yang, J., 2014. Effect of graphite zone in the formation of unconformity-related uranium deposits: insights from reactive mass transport modeling. *Journal of Geochemical Exploration,* DOI.org/ 10.1016/j.gexplo.2014.01.020.

Bethke, C. M., 1986. Hydrologic constraints on the genesis of the Upper Mississippi Valley mineral district from Illinois Basin brines. *Economic Geology*, vol. 81, 233-249.

Betts, P. G., Giles, D., and Lister, G. S., 2003. Tectonic environment of shale-hosted massive Sulfide Pb-Zn-Ag deposits of Proterozoic northeastern Australia. *Economic Geology*, vol. 98, 557-576.

Bierlein, F. P., and Betts, P. G., 2004.The Proterozoic Mount Isa Fault Zone, northeastern Australia: is it really a ca. 1.9 Ga terrane-bounding suture? *Erath and Planetary Science Letters*, vol. 225, 279-294.

Cui, T., Yang, J., and Samson, I. M., 2012a. Uranium transport across basement/cover interfaces by buoyancy-driven thermohaline convection: implications for the formation of unconformity-related uranium deposits. *American Journal of Science*, vol. 312, 994-1027.

Cui, T., Yang, J., and Samson, I. M., 2012b. Tectonic deformation and fluid flow: implications for the formation of unconformity-related uranium deposits. *Economic Geology*, vol. 107, 147-163.

Cuney, M., and Kyser, K., 2008. Recent and not-so-recent developments in uranium deposits and implications for exploration. Mineralogical Association of Canada (MAC) Short Course Series, vol. 39. Quebec, Canada. 215 pp.

Feltrin, L., McLellan, J. G., and Oliver, N. H. S., 2009. Modelling the giant, Zn-Pb-Ag Century deposit, Queensland, Australia. *Computers & Geosciences*, vol. 35, 108-133.

Frind, E. O., 1982. Simulation of long-term transient density-dependent transport in groundwater. *Advances in Water Resources*, vol. 5, 73-88.

Hanor, J. S., 1996. Controls on the solubilization of lead and zinc in basin brines. In: Sangster, D. F. (Ed.), Carbonate-hosted Lead-Zinc Deposits. *Economic Geology Special Publication*, vol. 4, 483-500.

Hobbs, B. E., Ord, A., Archibald, N. J., Walshe, J. L, Zhang, Y., Brown, M., and Zhao, C., 2000. Geodynamic modelling as an exploration tool. Australian Institute of Mining and Metallurgy Proceedings, Annual

Conference, After 2000-The Future of Mining, AusIMM Annual Conference Proceedings 2000, Melbourne, p. 34-39.

Ju, M., and Yang, J., 2011. Numerical modeling of couple fluid flow, heat transport and mechanicdeformation: example from the Chanziping ore district, south China. *Geoscience Frontiers*, vol. 2, 577-582.

Klein, S. A., and Harvey, A. H., 1996. NIST/ASME Steam Properties (version 2.0), Standard Reference Data Program. National Institute of Standard and Technology, Maryland, USA.

Leisman, H. M., and Frind, E. O., 1989. A symmetric-matrix time integration scheme for the Efficient solution of advection-dispersion problems. *Water Resources Research*, vol. 25, 1133-1139.

McLellan, J. G., Oliver, N. H. S., and Hobbs, B. E., 2006. The relative effects of deformation and thermal advection on fluid pathways in basin-related mineralization. *Journal of Geochemical Exploration*, vol. 89, 271-275.

O'dea, M. G., Lister, G. S., and MacCready, T., 1997. Geodynamic evolution of the Proterozoic Mount Isa terrain. *Geological Society of London Special Publication*, vol. 121, 99-122.

Oliver, N. H. S., McLellan, J. G., Hobbs, B. E., Cleverley, J. S., Ord, Al., and Feltrin, L., 2006. Numerical models of extensional deformation, heat transfer, and fluid flow across basement-cover interfaces during basin-related mineralization. *Economic Geology*, vol. 101, 1-31.

Schaubs, P. M., Rawling, T. J., Dugdale, L. J., and Wilson, C. J. L., 2006. Factors controlling the location of gold mineralisation around basalt domes in the Stawell corridor: insights from coupled 3D deformation-fluid-flow numerical models. *Australian Journal of Earth Sciences*, vol. 53, 841-862.

Yang, J., Latychev, K., and Edwards, R. N., 1998. Numerical computation of hydrothermal fluid circulation in fractured earth structures. *Geophysical Journal International*, vol. 135, 627-649.

Yang, J., Large, R., and Bull, S., 2004a. Factors controlling free thermal convection in faults in sedimentary basins: implications for the formation of zinc-lead mineral deposits. *Geofluids*, vol. 4, 237-247.

Yang, J., Bull, S., and Large, R., 2004b. Numerical investigation of salinity in controlling ore-forming fluid transport in sedimentary basins: Example of the HYC deposit, Northern Australia. *Mineralium Deposita*, vol. 39, 622-631.

Yang, J., 2006. Full 3-D numerical simulation of hydrothermal fluid flow in faulted sedimentary basins: example of the McArthur Basin, Northern Australia. *Journal of Geochemical Exploration*, vol. 89, 440-444.

Yang, J., Large, R., Bull, S., and Scott, D., 2006. Basin-scale numerical modelling to test the role of buoyancy driven fluid flow and heat transport in the formation of stratiform Zn-Pb-Ag deposits in the northern Mt Isa basin. *Economic Geology*, vol. 101, 1275-1292.

Yang, J., Feng, Z., Luo, X., and Chen, Y., 2009. On the role of buoyancy force in the ore genesis of SEDEX deposits: example from northern Australia. *Science in China*, vol. 52, 452-460.

Yang, J., Feng, Z., Luo, X., and Chen, Y., 2010. Three-dimensional numerical modeling of Salinity variations in driving basin-scale ore-forming fluid flow: Example from Mount Isa Basin, northern Australia. *Journal of Geochemical Exploration*, vol. 106, 236-243.

Yang, J., 2011. Fluid flow and heat transport: theory, numerical modeling and applications for the formation of mineral deposits. In: Berg, E. T. (Ed.), Fluid Transport: Theory, Dynamics and Applications. Nova Science Publishers, Inc., Chapter 3, p. 75-120.

Zhang, Y., Sorjonen-Ward, P., Ord, A., and Southgate, P., 2006. Fluid flow during deformation associated with structural closure of the Isa Superbasin after 1575 Ma in the central and northern Lawn Hill Platform, Northern Australia. *Economic Geology*, vol. 101, 1293-1312.

Zhang, Y., Roberts, P., and Murphy, B., 2010. Understanding regional structural controls on mineralization at the Century deposit: A numerical modelling approach. *Journal of Geochemical Exploration*, vol. 106, 244-250.

Zhao, C., Hobbs, B. E., Ord, A., Hornby, P., and Peng, S., 2008. Morphological evolution of three-dimensional chemical dissolution front in fluid-saturated porous media: a numerical simulation approach. *Geofluids*, vol. 8, 113-127.

In: Basins
Editor: Jianwen Yang

ISBN: 978-1-63117-510-7

Chapter 4

SUBSIDENCE, HYDROCARBON GENERATION AND THERMAL HISTORY MODELLING OF INLAND SEDIMENTARY BASINS (CHAD BASIN) IN NIGERIA

C. N. Nwankwo*[*1], A. S. Ekine[1], L. I. Nwosu[1] and M. Y. Kwaya[2]

[1]Department of Physics, University of Port Harcourt, Nigeria
[2]Department of Civil Engineering Technology, Federal Polytechnic, Damaturu, Nigeria

ABSTRACT

Subsidence, burial and thermal history studies are crucial factors in determining the hydrocarbon potential (both quality and quantity) of a basin. In this study subsidence and thermal analyses of Chad Basin Nigeria were investigated utilizing logs data from 23 exploratory wells. The oceanic lithospheric cooling concept was used for the basin subsidence while the sedimentation history in the basin was reconstructed utilizing the method of "back-stripping" at different locations. The results of the thermal and subsidence analyses were used to model the hydrocarbon maturation levels in different stratigraphic units of interest. Subsidence and sedimentation were observed to have occurred mostly in

* Corresponding author: Email: cyrilnn@yahoo.com.

the Albian to Maastrichtian times, and only the sediments from late Santonian units in the basin have reached sufficient thermal maturity (at a depth range of 1.5 to 3.8 km) to generate hydrocarbons. The hydrocarbon may still be occurring at present day. Maturation in terms of virtrinite reflectance (Ro) and maturation index (C) ranges from 0.6 and 1.37 percent, and 9.9 to 15.2, respectively, for these matured sediments. Time-temperature relation has been demonstrated to be a controlling factor for diagenetic and organic metamorphic processes in rocks. Level of maturation of organic matter was expressed using various maturation indices such as paleotemperature, oil window concept and virtrinite reflectance. The thermal model results show that basement subsidence is a function of the square root of the age of the basin, indicating that subsidence in Chad Basin Nigeria can be explained by a simple thermal contraction of the lithosphere following an extensional phase. With the results of the various modelling indices, the study has revealed the prospect of the basin to generate hydrocarbons in a commercial quantity.

Keywords: Subsidence, burial history, hydrocarbon generation, maturation, thermal modelling

INTRODUCTION

Basin modelling is the reconstruction of a basin history which includes the procedure of establishing the sequential records of changes that have occurred during the long geologic history of the basin. The objectives are to define the geometry and geochemical history of the basin, so as to determine the hydrocarbon potentiality of the basin.

The present day national petroleum reserves asset of about 38 billion barrels of oil and 190 trillion standard cubic feet of gas (Obaje et al., 2009) are derived solely from the onshore and offshore Niger Delta Basin. Some exploration campaigns have been undertaken in the inland basins with the aim of expanding the national exploration and production base and thereby add to the proven reserves asset. Such oil exploration efforts have not been successful as no meaningful oil or gas discovery of commercial quantity has been made in any of these basins. This may be attributed to a poor knowledge of the geology of the area and a lack of interest by exploration companies to invest in Nigerian Inland basins.

Hydrocarbons are formed by thermal alteration of organic sediments after burial through long span of time. One of the latest tools used in exploring the oil and gas potential of a basin is the determination of burial and thermal histories of the sedimentary rocks within the basin. Subsidence and thermal histories, when combined with seismic stratigraphic studies, help to determine the level of organic maturation, generation, and depth to oil generative window in the basin (Onuoha and Ekine, 1999; Ekine and Onuoha, 2008; Nwankwo et al., 2009; Emujakporue et al., 2009; Nwankwo and Ekine, 2009; Ekine and Onuoha, 2010). While thermal history modelling helps to quantify the hydrocarbons content and composition of a basin, the burial history is crucial in determining the generation, migration and preservation of hydrocarbons in the basin.

Wood (1988) and Henrickson and Chapman (2002) have demonstrated the importance of time – temperature relation in controlling diagenetic and organic metamorphic processes in rocks. Petroleum hydrocarbons are formed by thermal alterations of organic rich sediments during burial. Generation of hydrocarbons is therefore related to burial depths, time and temperature of source rocks (Nwankwo, 2007).

Previous works on Chad Basin Nigeria have centred mainly on interpreting the evolution of the basin from geological and structural evidences. Adepelumi et al. (2011) in characterizing the reservoir sand of the basin from well log data affirmed that the Net-to-Gross values indicate the presence of quality potential reservoir rocks. Obaje et al. (2006) in their geochemical study of the area recognized coal beds and Type III kerogen as the effective source rocks, and concluded that commercial prospects may exist in sedimentary basins of Northern Nigeria. Also, Omosanya et al. (2011) in evaluating the hydrocarbon potential of the basin from reservoir parameters indicated that Chad Basin holds considerable prospect for hydrocarbon in terms of reservoir abundance and estimated possible reserves. Obaje (2000) and Akande et al. (1998) in characterizing the basin source rock favoured Type III kerogen and a decrease in thermal maturity with depth, which implies an accumulation of gaseous hydrocarbon at deeper depths. Other works include Suh et al. (2000), Dike (1993), and Nwankwo et al. (2009, 2012).

Objective of the Study

The work is intended to provide an understanding on how well log petrophysical data can be utilized to evaluate hydrocarbon potential of a basin. Primarily, we intend to:

(i) Determine the maturation history of the sedimentary sequence in the basin.
(ii) Determine the type of hydrocarbon generated and expelled during basin evolution.
(iii) Increase our knowledge and understanding of the geology, petroleum prospect and potentials of the Chad Basin Nigeria.

Well Log measurements when properly calibrated can give petrophysical parameters necessary for qualitative and quantitative studies. The study results will be of benefit to oil industry operators, and also increase the Nigerian Government revenue since further drilling of dry wells in the basin will be minimized.

LOCATION AND ORIGIN OF CHAD BASIN

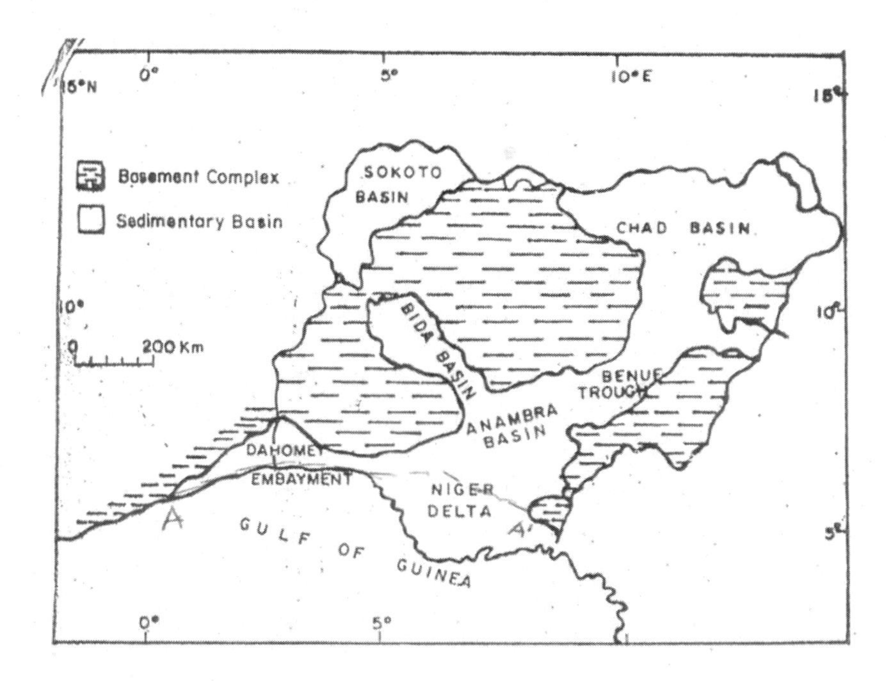

Figure 1. Sedimentary basins in Nigeria.

The Chad Basin lies within a vast area of Central and West Africa at an elevation of between 200 m and 530 m above sea level. The basin is the largest inland basin in Africa occupying an area of approximately 2,500, 000 km^2 extending over parts of the Republic of Niger, Chad, Sudan and the northern portions of Cameroon and Nigeria. The origin of the Chad Basin has been generally attributed to the rift system that developed in the early Cretaceous when the African and South American lithospheric plates separated and the Atlantic opened. Pre-Santonian Cretaceous sediments were deposited within the rift system.

The Nigerian sector of the Chad Basin (Fig. 1) is one of the Nigeria's inland basins, occupying the northeastern part of the country and covering Bornu State and parts of Yobe and Jigawa States (Obaje et al., 2011; Adepelumi et al., 2011). It represents about one-tenth of the total area of the Chad Basin, and falls between Latitudes 11°N and 14°N and Longitudes 9°E and 14°E. The altitude of the basin ranges from 300 m within the lake to about 530 m at the western margin, along a distance of about 240 km. The Chad Basin has been referred to as an interior sag basin, and has developed at the intersection of many rifts, mainly in an extension of the Benue Trough. Major grabens then developed and sedimentation started.

TECTONIC SETTING

The Chad Basin presents a rather complex tectonic history that is still not fully understood. The basin is an intracratonic depression of the Pre-Cambrian basement associated with the early Cretaceous rift extensions of the Benue trough Complex. Genik (1992) presented a model for the regional framework and tectonic evolution of the Cretaceous-Paleogene rift basins of Niger, Chad and the Central African Republic. Both geophysical and geological interpretations of data suggest a complex series of Cretaceous grabens extending from the Benue trough to the southwest. These data imply a similar tectonic origin involving crustal thinning within and beneath the grabens and the near-surface presence of igneous intrusions within the horst/graben structures overlain by a relatively thick succession of sedimentary materials. There is a preponderance of tensionally induced basement tectonics and complex pattern of faulting that led Avbovbo et al. (1986) to classify the Chad Basin as a rift related basin. There was a reduction in rift development at the late rift stage of the basin which according to Genik (1992) was due to changes in the stress regime caused by variations in the spreading rate and direction between the Equatorial and South Atlantic plates (Fairhead and Blinks, 1991), culminating in a marked tectonic pulse of about 85 Ma.

Most faults in the Nigerian sector of the Chad Basin are basement-induced. Most induced-faulting gives rise to horst and graben features, and the movement along such faults are translated into high angle fault systems in the overlying sediments. Detached faults probably developed in response to the basinal deformation (Sag phase) and show an overall increase over the basement controlled faulting. Throws along these faults vary in magnitude with most of them characterized by a decrease with depth (Avbovbo et al., 1986).

Folds within the basin are simple and symmetrical structures with low fold frequencies and amplitudes, which increase towards the center of the basin (Okosun, 2000), and have a spatially restricted occurrence to the southern part (Avbovbo et al., 1986). The folds are considered flexural folds, which developed as a result of thermal subsidence in the basement-controlled graben along the fault/fracture planes. The frequency decreases northwestwards towards the shallow part of the basin until they die out as minor ripples.

Although, there is no direct evidence of pre-Cretaceous rocks in the Nigerian sector of the Chad Basin, it is believed that these sediments may be preserved in the lower depressions and grabens which characterise the basin floor topography.

BASIN STRATIGRAPHY

Sedimentary rocks are mainly continental, sparsely fossiliferous, poorly sorted, and medium to coarse-grained, feldspathic sandstones called the Bima Sandstone. A transitional calcerous deposit – Gongila Formation that accompanied the onset of marine incursions into the basin, overlies the Bima Sandstones. These are overlain by graptolitic Fika shale. The Kerri-Kerri Formation of Paleocene age, which represents an unconformable continental sequence of flat lying grits, sandstone and clays make contact with the marine Fika Shale at its basal portion. The Chad Formation with a total thickness ranging from 300 m to 1200 m is the youngest stratigraphic unit in the basin.

Stratigraphic descriptions of the southern Chad Basin (Nigerian sector) are available in Okosun (2000), and comprise Mesozoic to Quaternary sediments. Ejedawe et al. (1985) have identified two stratigraphic breaks of some significance in the evolutionary history of the basin. Hydrocarbons could be trapped by structural or stratigraphic means. Ayoola et al. (1982) pointed out that stratigraphic traps are provided by off lap sedimentary features and unconformities.

MATERIALS AND METHOD

Basic geophysical and lithologic logs of twenty three oil exploratory wells in the study area were provided by the Nigerian National Petroleum Corporation through its frontier exploration services arm, National Petroleum Investment Management Services (NAPIMS) Maiduguri. Geothermal gradients were computed from corrected bottom hole temperatures while the bulk effective thermal conductivity for the different stratigraphic units encountered in the wells were computed from the sonic logs. The heat flow values were then calculated, and the geothermal gradient and heat flow maps for the basin were constructed (Nwankwo et al., 2009). The thermal subsidence of the basin was computed from lithospheric isostatic model, while the burial history was reconstructed using 'backstripping' or decompaction method. Utilising the computed geothermal gradient, thermal conductivity and heat flow estimates for the basinal units, the maturation history of the source rocks and evolution of the oil generative window (OGW) in the 23 oil wells were evaluated. Timing of hydrocarbon generation was deduced from the maturation history using a correlation between calculated maturity and virtrinite reflectance. The type of hydrocarbon generated within the OGW was inferred from the thermal conditions and maturity levels.

THERMAL SUBSIDENCE

The basin subsidence is assumed to be an isostatic response to the thinning of crust and cooling of a thermal anomaly. As time progresses additional load resulting from replacement of seawater by the accumulating sediments increases the subsidence rate.

The thermal subsidence of the basin was modelled by extending the cooling model for the oceanic lithosphere to sedimentary basins. Ekine and Onuoha (2008) and Onuoha (1985) expressed the depth to basement of the sedimentary basin as a result of the thermal decay, which is the thermal subsidence, as:

$$Z_{SB} = \frac{2\rho_m \alpha_m (T_m - T_o)}{\rho_m - \rho_S} \left(\frac{K_m t}{\pi} \right)^{1/2} \tag{1}$$

where Z_{SB} is thickness of the sediment, ρ_S is average density of sediment, ρ_m is density of mantle, T_o is surface temperature, t is age of sediment in million years, α_m is coefficient of thermal expansion of mantle, and K_m is thermal diffusivity of mantle.

The depth to sediment, Z_S (t) deposited at time t_s after the initiation of subsidence is given by the expression:

$$Z_S(t) = E_O\left(t^{1/2} - t_S^{1/2}\right) \tag{2}$$

where

$$E_O = \frac{2\rho_m\alpha_m(T_m - T_o)}{\rho_m - \rho_S}\left(\frac{K_m}{\pi}\right)^{1/2}$$

and t_s is time when sediment was deposited after the initiation of subsidence.

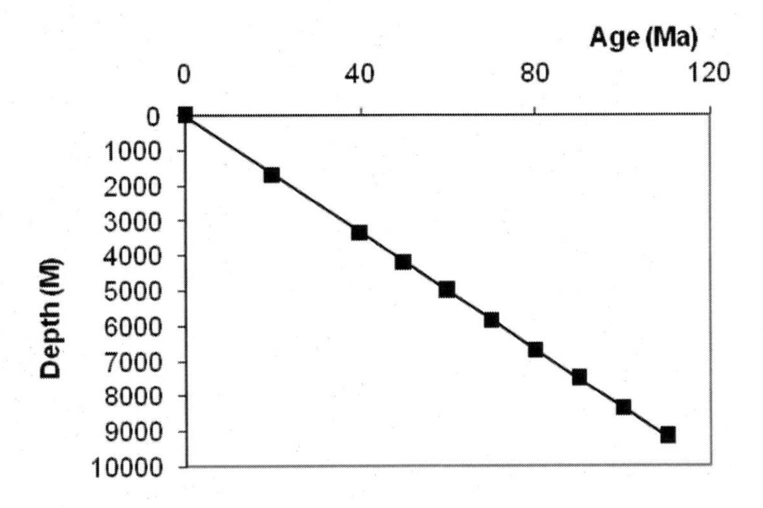

Figure 2. Subsidence history of Chad Basin Nigeria.

Figure 2 shows the dependence of depths to the basement as a function of age using equation 1. The plot was constructed by utilizing the physical constants listed in Table 1. Equation (2) was used to predict the depths to several sedimentary layers deposited t_S million years after subsidence began (Fig. 3). Initiation of subsidence is taken as the time of deposition of the oldest sediment in the basin, which is the Bima Sandstone in the Upper Albian (110 Ma). The mean saturated sediment density for the basin was taken as 2220 kgm^{-3}. All other constants used in the computation are listed in Table 1. The effect of water depth and sea level changes have not been considered in calculating the depth to basement in this study because of the observations by Onuoha and Ofoegbu (1988) that the corrections for these changes are usually very small such that the inaccuracies in determining the paleowater depth may be greater than the correction itself.

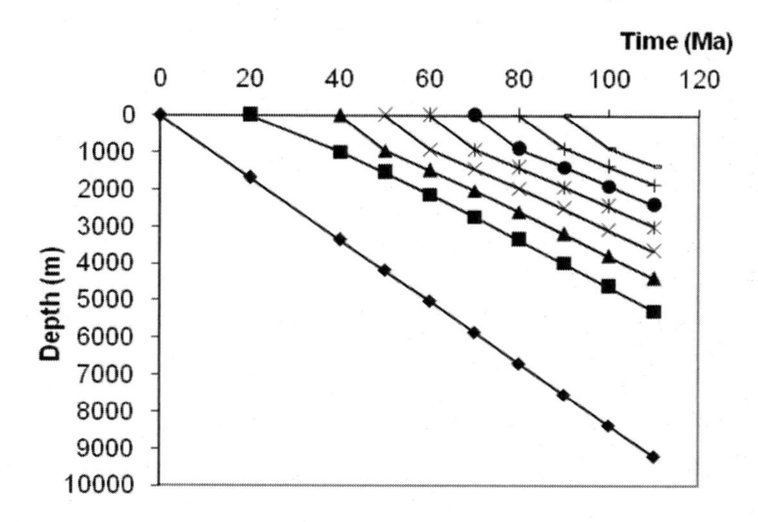

Figure 3. Depths to sediment deposited at times t_s as a function of time.

Table 1. Physical constants used in various computations

Mantle density, ρ_m	3380kgm^{-3}
Temperature of mantle , T_m	1300^0 C
Surface temperature, T_0	27^0 C
Coefficient of thermal expansion of mantle α_m	2 x 10^{-6} $^0C^{-1}$
Thermal diffusivity of mantle K_m	10^{-7} m^2 s^{-1}

Burial History and Sediment Decompaction

Sedimentation is an active process occurring at an approximate rate of 1.0 mm/y or 1km/Ma (Alexander and Flemings, 1995). With the exception of salt, all sedimentary sequences compact and dewater on burial, with shale and coal undergoing stronger compaction relative to other lithologic units. This compaction increases the bulk density of the sedimentary layers and reduces their porosity.

Reconstructing the sedimentation history of passive margins and sedimentary basins helps in assessing the hydrocarbon potentials of the basin. Properly constructed sedimentation or burial history diagrams illustrate changes in the column of sediments above the initial layer. To reconstruct the burial history of a sedimentary basin by 'back-stripping' or decompaction method requires the establishment of porosity-depth relation in the basin. A decompaction and backstripping method similar to that described by Sclater and Christie (1980) was used in this study. A generalized porosity-depth approximation exhibited by variety of sediments is given by

$$\varphi_z = \varphi_o e^{-cz} \qquad (3)$$

where φ_o is initial or depositional porosity, c is a constant, the depth coefficient and z is the depth. Porosities, φ_z, were computed from interval transit times obtained from sonic logs at various depths using the empirical relation:

$$\varphi = \frac{\Delta t - \Delta t_m}{\Delta t_f - \Delta t_m} \qquad (4)$$

where Δt is acoustic travel time at the point of consideration in μsec/m, Δt_m is acoustic travel time of the rock matrix in μsec/m, and Δt_f is acoustic travel time of interstitial fluid (water) in μsec/m.

Assuming the above porosity behaviour (equation 3), we strip off the top of individual stratigraphic units one at a time successively, such that the geology of earlier times is restored. Following Ekine and Onuoha (2008); Sclater and Christie (1980); Watts and Ryan, 1976; if the top of the unit is z_1 and the bottom of the unit is z_2 (Fig. 4), the depth to the top of a unit z_1' and that to the bottom of the unit z_2' of an earlier time is given by

$$z_2' = h_{sg} + z_1' - \frac{\varphi_0}{c}\left[e^{-cz_1'} - e^{-cz_2'}\right] \qquad (5)$$

where

$$hsg = z_2 - z_1 - \frac{\varphi_0}{c}[e^{-cz_1} - e^{-cz_2}] \quad (6)$$

is the height of the sediment grain for a unit cross-sectional area between the depth intervals z_1 and z_2.

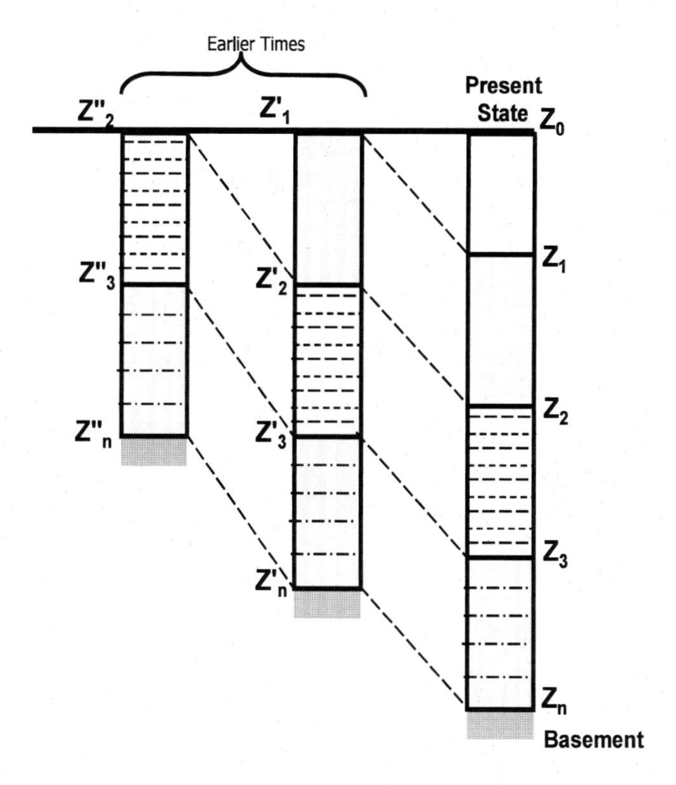

Figure 4. Schematic diagram for the decompaction of sedimentary layers.

We solve equation (5) for z_2' by setting z_1' to zero since it is the bottom of the previous layer. Subsequent units (all subscripts in equation (5) are increased in steps of one) are solved iteratively in the same manner until the basement or the assumed base is at the surface. An average initial porosity of 67% and a depth coefficient of 0.4077 have been estimated for the basin using equation 3, from porosity-depth variation analysis and sonic log data

(Nwankwo et al., 2009). Also, the basement of the basin has been assumed to be 110 Ma.

The burial history curve for Murshe-1 Well, which is representative of all the 23 Well locations analyzed, is represented in Fig. 5. The gradients of the burial curves, which are measures of rates of the basin subsidence at any time, generally decrease with time, and this implies that the rate of subsidence decreases with age. A rapid burial rate is noticed from the curves to have occurred within the Miocene Units. This period must also have been characterized with upliftment, erosion and subsidence.

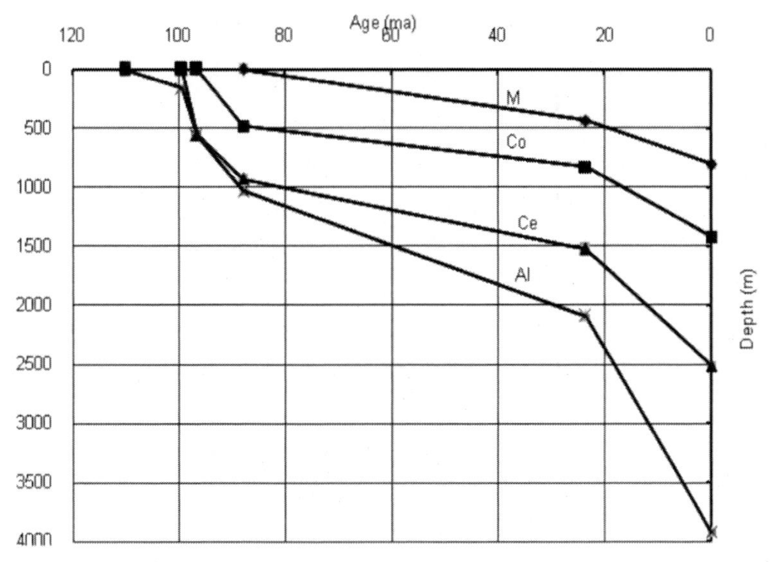

M=Miocene Co=Coniacian Ce=Cenomanian Al=Albian

Figure 5. Burial history of Murshe-1 Well.

THERMAL AND MATURATION HISTORY

Hydrocarbon generation from thermal reactive organic materials during burial is part of the overall process of thermal metamorphism of organic matter. The measured maturity values for possible source rocks are a vital tool in knowing the present-day maturity levels and hydrocarbon generation status.

Hydrocarbon maturation and generation trends were modelled based on the thermal maturation profiles for oil wells drilled in the basin. The thermal history of the sequence in each location is related to the change in geothermal gradient and burial rates with time. The subsidence and thermal history reconstructions help to identify the timing of dominant episodes of heating and cooling that have affected the basin. These computed parameters were then used in modelling the thermal maturity and hydrocarbon potential of the entire basin. Among the various techniques usually employed in calculating sediment maturity levels, only the oil window, paleotemperature and virtrinite reflectance concepts have been considered in this study.

THE OIL WINDOW CONCEPT

The source rock maturity of a sedimentary basin can be assessed based on the average depth to the oil floor calculations (Piggot, 1985). Below the diagenetic zone of buried sediment, the oil ceiling, or depth of intense oil generation, is defined as the depth below which oil generation begins to increase substantially, while the depth and associated temperature at which oil is no longer generated and gas begins to dominate (metagenesis zone) is known as the oil floor. These two depths bound the oil generative window (zone of catagenesis). An oil window is therefore the space-time continuum inside which liquid hydrocarbons are generated and preserved.

Piggot (1985) expressed the time (in million years) for maturation of an organic matter as

$$t=1.910\exp(-0.0408T) \qquad (7)$$

where T is the oil threshold temperature in ^{o}C determined from Arrhenius relationship as a function of sediment age. It is expressed as:

$$T = 164.4 - 19.39 \ln t \qquad (8)$$

Depths for the oil ceiling D_{oc} is given by

$$D_{oc} = 100(T - T_s)/(dT/dZ) \qquad (9)$$

where T_s is the mean surface temperature, and dT/dZ is the geothermal gradient in ^{o}C/100 m.

Similarly, the depths to oil floor is calculated from the relationship

$$D_{of} = 100(150 - T_s)/(dT/dZ) \tag{10}$$

and

$$\text{Oil window} = D_{of} - D_{oc} \tag{11}$$

Equation (11) has been used to estimate the 'oil window' for the wells in the Chad Basin. In estimating the oil window for the basin, the age of the top Fika Shale, the potential source rock in Chad Basin is taken to be Top Maastrichtian (66.5 Ma) (Kent and Gradstein, 1985). Inserting this value into equation (8) gives the oil threshold temperature, T, for the basin to be 83 °C.

PALEOTEMPERATURE AND VIRTRINITE REFLECTANCE

Measuring only the bottom hole temperature from oil well boreholes does not answer the question of sediment maturity as this only indicates the present day temperature, which may be considerably lower than that of the past, especially in areas that have undergone large-scale uplift and erosion. The important relationship between temperature, petroleum generations and heat flow in sedimentary rocks during burial makes it imperative to determine the paleotemperature and hence the thermal potential for maturation of such strata. Hydrocarbons are formed by the thermal alteration of organic-rich sediments during burial, and the process is primarily dependent on the integrated time/temperature history of the buried organic material.

Several properties like porosity, permeability, fluid and kerogen type in source rocks, particle size and distribution, pressure gradient, viscosity, and saturation significantly influence the generation and migration of petroleum. In order to predict the hydrocarbon history of a basin and to fully establish its maturity, we must determine the temperature as a function of time, and know the variation of heat flow or geothermal gradient with time, the thickness of each sedimentary layer as a function of time, and thermal conductivities as function of time and depth.

Sclater and Christie (1980) determined a near surface temperature at time t and depth z by using the expression:

$$T(t,z) = T_{surface} + \int_{o}^{z} \frac{Q(t)dz}{K(t,z)} \tag{12}$$

where K(t,z) is the conductivity as a function of time and depth, $T_{sufarce}$ is the surface temperature and Q(t) is the heat flow. Assuming a surface temperature of 27°C and utilizing the conductivity and heat flow values calculated for the basin (Nwankwo et al., 2009; 2010), equation (12) was numerically integrated for each sedimentary layer.

In calculating the degree of thermal maturation of a given sedimentary basin, Royden et al. (1980) established the hydrocarbon thermal alteration parameter C to be related to temperature by the relation:

$$C = \ln \int_0^t 2^{T(t,z)/10} \, dt \tag{13}$$

where T(t,z) is the temperature as a function of time t. This relationship is based on the observation that the reaction rate for thermal alteration of organic sediments doubles for every 10 °C increase in temperature. Uko (1996) has shown to a first approximation that the oil generating process has hardly started when C is equal to 10 and is essentially complete when it is greater than 16. Most light oils are expelled when C lies between 12 and 14, and gas generation is complete when C is greater than 20.

The index C is also related to the level of organic maturity and virtrinite reflectance Ro. Reflectivity is generally accepted as a measure of thermal alteration of organic matter, and is assumed to increase with temperature and time. It is given as

$$\log (Ro) = a + bC \tag{14}$$

where a = 0.995 and b = 0.079.

Combining equations 13 and 14, Middleton (1982) expressed the virtrinite reflectance as:

$$(R_o)^{5.5} = 3.4 \times 10^{-6} \int_0^t \exp[0.069\, T(t)] \, dt \tag{15}$$

where R_o is expressed in percentage (%), T in °C and t in million years (Ma). Crude oil generation occurs for reflectivity R_o values between 0.6% and 1.5%. Gas generation takes place for R_o between 1.5% and 3.0%; and at values greater than 3.0%, rocks are graphitic without any hydrocarbons. An abrupt shift in the value of R_o with depth may indicate faults or unconformities, while an abrupt increase in R_o with depth followed by a return to the previous gradient may be caused by igneous intrusive. Basins that are not largely affected by major unconformities, faulting and localized igneous activity have a linear relationship between depths and log R_o.

Equations 12, 13 and 15 have been evaluated in order to predict the maturation conditions in the Chad Basin, Nigeria. Figures 6, 7 and 8 show the thermal history for the sedimentary layers in the Kanadi-1 Well that is representative of all the Well locations. Temperature increases with depth and decreases with age, with the older sediments having been heated more slowly in the recent past. Most late Cretaceous and younger sediments in the basin have temperatures below 80 °C.

Virtrinite reflectance Ro has become one of the most widely used indicators of maturity of organic materials. It varies with temperature and time, and the slope of Ro versus Depth indicates the geothermal gradients in the history of the basin. There is an increasing trend in reflectance with depth and age.

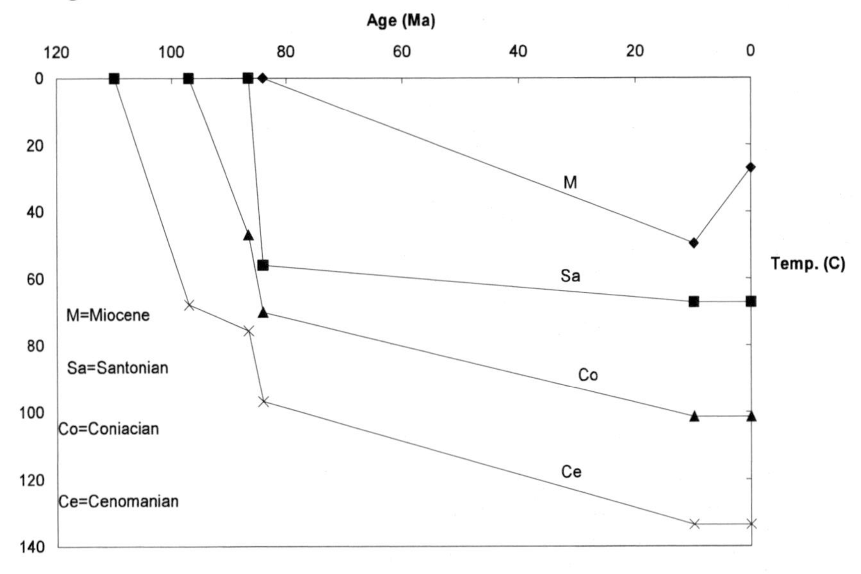

Figure 6. Temperature history for Kanadi-1 well.

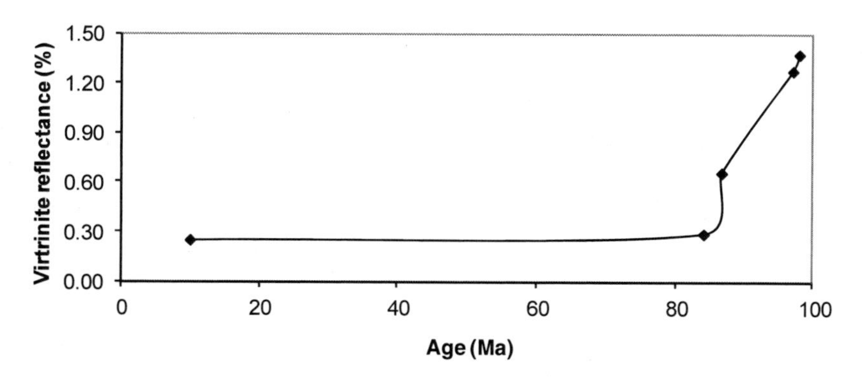

Figure 7. Age versus virtrinite reflectance variation for Kanadi-1 Well.

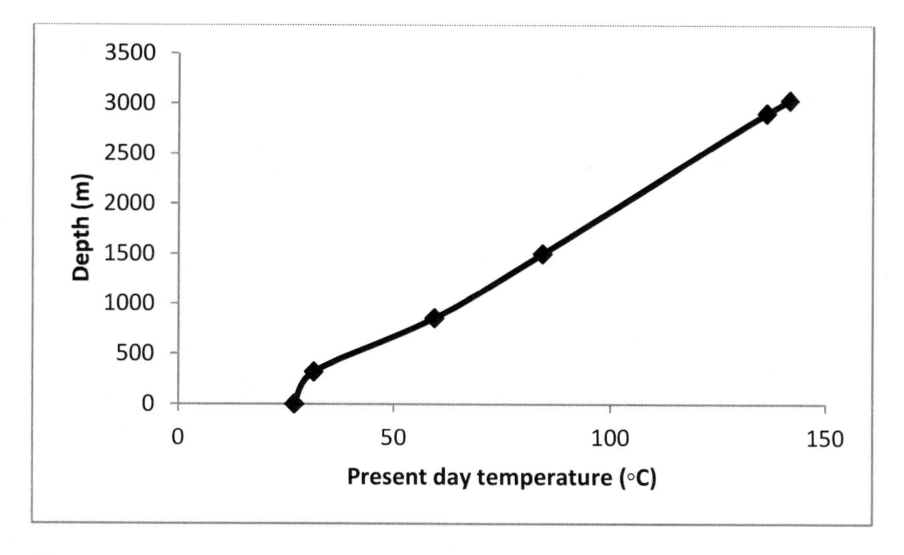

Figure 8. Depth versus temperature variation for Kanadi-1 Well.

Discussion of Results

The subsidence history of the Chad Basin has been reconstructed in an attempt to explain the origin and subsequent evolution of the basin. Initiation of subsidence is taken as the time of deposition of the oldest sediment which is of late Albian age (about 110 Ma). The plot of basement subsidence as a function of the square root of the age of the basin (Fig. 9) indicates that the origin of subsidence in Chad Basin can be explained by a simple thermal contraction of the lithosphere following an extensional phase. Any subsequent evolution of the basin must have been influenced by some tectonic movements and weight of accumulating sediments.

The results of thermal history analysis reveal a declining burial rate with time, interspersed with two periods of unconformity development. From the Age/Temperature/Depth plots, it is observed that the post-Santonian sediments for most oil wells in the Chad Basin have hardly been subjected to temperatures higher than 80 °C at any time. This means that these sediments are immature and hence have poor hydrocarbon prospects. However, the potential source rocks from late Santonian units in the basin have reached sufficient thermal maturity to generate hydrocarbons and that generation may still be occurring at present day. There was an early history of rapid temperature build-up, followed by a period of slow temperature build-up, indicating a general cooling of the basin.

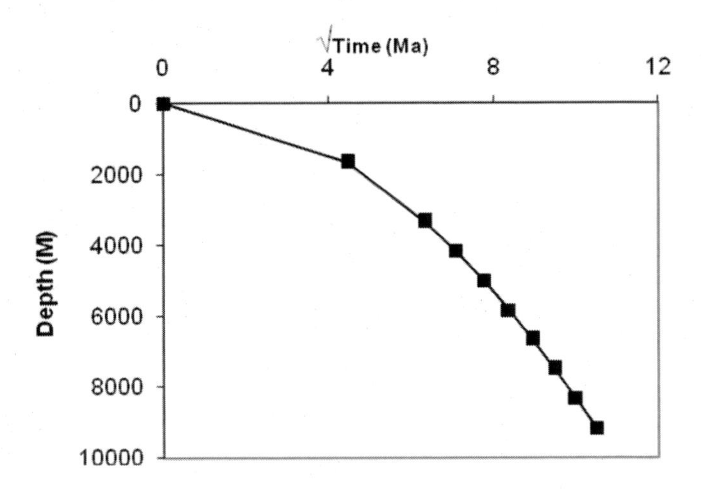

Figure 9. Subsidence versus √time.

One important element in the preservation of any generated hydrocarbons is the past and present temperatures of the various lithostratigraphic units. This study reveals that Fika-Gongila Formations are at temperature range of 60 °C to 158 °C while the Bima Formation is presently at temperatures of 102 °C to over 190 °C. At the high temperatures of the Bima Formation only methane gas will be preserved, while the Fika-Gongila would still preserve some oil, wet gas and condensate. The shallower intervals of Kerri-Kerri Formation are relatively not thermally mature.

The uppermost section of the Fika Shale has been found to have low levels of thermal evolution, and the volcanic intrusives may have invaded most of the Cretaceous section, which may have affected the quality of the source and reservoir rocks. Obaje et al. (2004) have reported that the source rock quality for Benue Trough and Chad Basin sediments is fair to poor.

Also, the Fika-Kerri-Kerri and the Kerri-Kerri-Chad erosional unconformities may account for substantial losses of some of the Fika and suprajacent sediments of Pliocene to Miocene ages. The sediments maturity (oil generation process) in most of the wells seems to have started and completed between the depths of 1.5 km and 3.8 km. Generally, an oil window range of 1.27 km and 4.10 km, and an average depth to oil floor of 3.63 km were estimated using oil window concept. Presently, only four out of the twenty-three wells drilled in the basin have penetrated up to a total depth of 3.8 km. Above the oil window any potential source rock would not have generated liquid hydrocarbon; while below the window liquid hydrocarbons would have been converted to gaseous hydrocarbons. In general, the oil formation is stronger in young source rocks due to a high geothermal gradient and oil can form at shallow depths.

Ro values vary between 0.60 and 1.37 within the sediments' oil generative window, while it rises up to 1.58 in Herwa-1 well. This range correlates with a paleotemperature range of 77.7 to 146.3 ^{o}C, and 176.3 ^{o}C respectively, which is within the oil zone of the oil generative window. For age intervals where the sediments were deemed immature, Ro values range between 0.21 and 0.59. Correspondingly, the maturation index C varies between 9.9 and 15.2 where the sediments are mature and have the potential of generating oil. The onset of the thermal maturation is close to 80 ^{o}C when a C value of 10 is reached.

The fact that hydrocarbon discoveries have been made in wells drilled in Chad Republic, whose sediments type and age compares favourably with those of the Nigerian sector, could be that the Chadian environment favours shallow to deep levels of hydrocarbon occurrence. Also most values of the thermal maturity index calculated within the oil generative window in this study lie between 12 and 14. It is therefore possible that the oils in the basin are light and hence must have been expelled.

CONCLUSION

The burial and thermal histories of the basin are directly linked to the tectonic and depositional histories of the basin. The thermal history modeling indicates that effective petroleum source rocks are the Upper Cretaceous (Santonian-Turonian) shales. Tertiary shales, which may be high in organic content, have not been subjected to favourable burial and thermal histories required for petroleum generation.

Results of various maturation parameters obtained in this study are consistent, and reassure the hope for future discoveries of oil in commercial quantity in the basin if all necessary exploration precautions are given due considerations.

The fact that oil or gas is yet to be found in commercial quantity in the Nigerian sector of the Chad Basin does not rule out the possibility of having oil/gas reservoirs in this basin. There is no doubt that the Nigerian inland basins as a whole have not been highly explored probably because of the poor knowledge of their geology, and also due to the abundance of oil in the Niger Delta.

ACKNOWLEDGMENTS

The authors are grateful to Nigerian National Petroleum Corporation through its frontier exploration services arm, National Petroleum Investment Management Services Maiduguri, for providing the well logs data, and R & D (EXRS) (Mr. A. S. Martyns-Yellowe), NNPC, Port Harcourt for providing other materials relevant to this work. Similarly, the materials and advice provided by Dr. A.A. Zarma and Prof. I.B. Goni, Department of Geology, University of Maiduguri, Prof. J.O. Ebeniro, Department of Physics, University of Port Harcourt, and Dr. E.D. Uko, Department of Physics, Rivers State University of Science and Technology Port Harcourt are well appreciated.

REFERENCES

Adepelumi, A.A, Alao, O.A and Kutemi, T.F (2011). Reservoir characterization and evaluation of depositional trend of the Gombe sandstone, southern Chad basin Nigeria. *Jour. of Petroleum and Gas Engineering* 2(6): 118-131.

Akande, S.O, Ojo, O.J and Erdtmann, B.D (1998). *Thermal maturity of Cretaceous sedimentary successions in the Benue rift basins, Nigeria.* Abstract NMGS Conf. Ife 98, 12.

Alexander, L.L., and Flemings, P.B. (1995). Geologic evolution of a Plio-Pliestocene Salt withdrawal Mini-Basin; Block 330, Eugene Island, South Addition, offshore Louisiana, *AAPG Bull.*, 79 (12), 1737-1756.

Avbovbo, A.A., Ayoola, E.O and Osahon, S.A (1986). Depositional and structural styles in Chad basin of Northeastern Nigeria. *AAPG Bull.* Vol. 70, 121. 1787 – 1798.

Ayoola E.O, Amaechi, M and Chukwu, R (1982). Nigeria's Newer Petroleum Exploration Pronvinces, Benue, Chad and Sokoto Basins. *Journ. Min. Geol.* 19, (1) 72-87.

Dike E.F.C (1993). Stratigraphy and structure of the Kerri-Kerri basin, NE Nigeria. *Jour. Mining and Geol.* 29, 77-94.

Ejedawe, J.E. , Coker, J.L., Lambert- Aikhiowhare, D.O (1985). Evolution of the oil Generative window (OGW) in sedimentary basins. *NAPE continuing Education Course series 1.*

Ekine, A.S and Onuoha, K.M. 2008. Burial history analysis and subsidence in the Anambra Basin, Nigeria. *Nigerian jour. of Phy.* 20(1): 145-154.

Ekine, A.S and Onuoha, K.M (2010). Seiesmic geohistory and differential interformational velocity analysis in the Anambra Basin, Nigeria. *Earth Sci. Res. Jourr.* Vol 14, No. 1, 88-99.

Emujakporue, G. O., Ekine, A. S and Nwankwo, C. N (2009). Evaluation of the hydrocarbon maturation level in oil wells in sedimentary basin of the Northern Niger Delta, Nigeria. *Journ. Appl. Sciences and Environmental Management.* Vol. 13, No. 3, pg. 79 – 82.

Fairhead, J.D and Blinks, R.M (1991). Differential opening of the Central and South Atlantic Oceans and the opening of the Central African rift system, *Tectonophysics*, 187. 191-203.

Genik, G.J (1992). Regional Framework Structure and petroleum aspects of the rift basins in Niger, Chad and Central African Republic (C.A.R.), *Tectonophysics*, 213, 169-185.

Kent, D.V and Grandstein (1985). A Cretaceous and Jurassic geochronology: *Geol. Soc Amer. Bull.*, 96, 1419-1427.

Middleton, M.F (1982).The Subsidence and thermal history of the Bass Basin, Southeastern Australia. *Tectonophysics*, 87: 383-343.

Nwankwo, C.N (2007). *Heat flow studies and hydrocarbon maturation modelling in the Chad Basin, Nigeria.* Unpublished Ph.D dissertation, University of Port Harcourt.

Nwankwo, C.N.; A.S. Ekine and Nwosu L.I (2009). Estimation of the Heat Flow Variation in the Chad Basin, Nigeria. *Journ. Appl. Sciences and Environmental Management*. Vol. 13, No. 1, pg. 73 – 80.

Nwankwo, C.N. and Ekine, A.S (2009). Geothermal gradients in the Chad Basin, Nigeria, from bottom hole temperature logs. *Int. journal of physical sciences*. Vol. 4, No. 12, pg. 777 – 783

Nwankwo, C.N., Ekine, A.S and Nwosu, L.I (2010). *Geothermal studies of Chad Basin Nigeria: Implications to hydrocarbon potential.* LAP LAMBERT Academic Publishing GmbH & Co. KG 2010 Germany. ISBN: 978-3-8433-6696-0.

Nwankwo, C.N., Emujakporue, G.O. and Nwosu, L.I. (2012). Evaluation of the Petroleum Potentials and Prospect of the Chad Basin Nigeria from Heat Flow and Gravity Data. *Journ. of Petroleum Exploration & Production Technology* Vol. 2, pg. 1-5, DOI: 10.1007/s 13202-011-0015-5.

Obaje N.G (2009). *Geology and Mineral Resources of Nigeria*. Springer, Heidelberg, pp 221.

Obaje, N.G, Attah, D.O., opeloye, S.A and Moumouni, A (2006). Geochemical evaluation of the hydrocarbon prospects of sedimentary basins in Northern Nigeria. *Geochemical Jour.* vol 40, 227-243.

Obaje, N.G, Musa, M.K, Odoma, A.N and Hamza, A (2011). The Bida Basin in North-Central Nigeria: Sedimentology and Petroleum geology. *Jour. of Petroleum and Gas Explo. Research.* Vol. 1(1), 001-013.

Obaje, N.G., Wehner, H., Scheeder, G., Abubakar, M.B., and Jauro, A (2004). Hydrocarbon prospectivity of Nigeria's inland basins: From the viewpoint of organic geochemistry and organic petrology. *AAPG Bulletin*, 88, 3, 325-353.

Okosun, E.A (2000). A preliminary assessment of the petroleum potentials from southwest Chad Basin (Nigeria). *Borno journal of Geology* vol. 2/2, 40-50.

Omosanya, K.O, Akinmosin, A.A and Adio, N.A (2011). The hydrocarbon potential of the Nigerian Chad Basin from wireline logs. *Indian jour. of sci. and tech.* Vol 4, no. 12, 1668-1675.

Onuoha, K.M (1985). Basin subsidence, sediment decompaction and burial history Modeling techniques. Applicability to the Anambra basin. *NAPE proc.*, 2: 6-17.

Onuoha, K.M and Ekine, A.S (1999). Subsurface temperature variations and heat flow in the Anambra basin, Nigeria. *J. of African Earth Sciences*, 28, 3. 641-652.

Onuoha, K.M and Ofoegbu, C.O (1988). Subsidence and evolution of Nigeria's continental margin: Implications of data from Afowo-1 well. *Marine and Petroleum Geology,* 5: 175-181.

Pigott, J.D (1985). Assessment of source rock maturity in Frontier basin: Importance of Time, Temperature and Tectonics, *AAPG Bull.* V 69, no. 8, 1269-1274.

Royden, L., Sclater, J.G. and Von Herzen, R.P (1980). Continental margin heat flow: Important parameters in formation of petroleum hydrocarbons, *AAPG bull.,* 64, 173 – 187.

Sclater, J. G. and Christie, P.A.F (1980). Continental stretching: an explanation of the Post-Mio-Cretaceous subsidence of the central North Sea Basin, *Journ. of Geophy. Res.* Vol. 80, 3711 – 3739.

Suh, C.E, Shemang, E.M, Dada, S.S, Samalia, N.K and Likkason, O.K (2000). Brittle events at the Southern margin of the Kerri-Kerri Basin NE Nigeria: Evidence for a post paleocene extensional regime in the Upper Benue trough. *NAPE bull.* 15; 1-15.

Uko, E.D (1996). *Thermal modeling of the Northern Niger delta.* Unpublished Ph.D Thesis, University of Science and Technology, Port Harcourt.

Watts AB and Ryan WBF (1976). Flexure of the lithosphere and continental margin basins. Tectonophysics, 36, 25-44.

Wood, A.D (1988). Relationships between thermal maturity indices calculated using Arrhnius equation and Lopatin method: *AAPG Buletin*, 72, 2, 115-134.

In: Basins
Editor: Jianwen Yang
ISBN: 978-1-63117-510-7

Chapter 5

THE CONCEPT OF A BASIN IN LARGE FLATLANDS: A CASE STUDY OF THE NORTHWEST REGION OF THE BUENOS AIRES PROVINCE, ARGENTINA

Eduardo E. Kruse and Jorge L. Pousa*
Consejo Nacional de Investigaciones Científicas y Técnicas (CONICET), Argentina, Facultad de Ciencias Naturales y Museo, Universidad Nacional de La Plata, La Plata, Argentina

ABSTRACT

A good deal of the Earth's population inhabits large flatlands that very often supply huge quantities of food to humankind. Topographic slope of the hydrologic system of a plain is very small, or zero, which is sometimes insufficient to generate an integrated drainage network, a surface runoff or to cause fluvial erosive processes. A sector of the Pampas Plain, locally known as "Pampa Arenosa" (Sandy Pampa), in the northwest of the Buenos Aires Province, Argentina, is presented as a case study. With an area of about 50,000 km2, this region presents a monotonous plain surface of low relief. It is a landscape shaped by eolic processes where it is possible to distinguish different paleoforms related to dune ridges. Current wet climate conditions contrast with the origin of an eolic landscape associated with a preceding arid climate. The

* Corresponding author: E-mail: jlp@fcnym.unlp.edu.ar.

dominant features are the absence of a drainage network and the accumulation of water in blowouts or in inter dune openings. The general characteristics of the region, as well as the hydrologic factors (rainfall, water surplus, soil wetness, water table level) are analyzed, which reveals the most marked impacts from alternate flood and dry periods. The way in which the regional hydrologic variations affect the agricultural production system in one of the most important agricultural regions of Argentina is also studied. The main feature of a large flatland is its low morphogenetic energy, which makes the energy of water resources dissipate through vertical water movements (evaporation, transpiration, infiltration, interchanges between the unsaturated zone and the water table). These vertical movements predominate over surface or groundwater horizontal movements, thus enhancing the importance of surface and groundwater storage variations. It follows that flatlands are highly sensitive to climatic fluctuations (water surplus and deficit), as well as to human activities. These characteristics make it difficult to effectively apply the traditional concept of a basin used for regions having an abrupt relief. Moreover, the difficulty to define a study area as a drainage basin leads to a need for replacing the term drainage basin by hydrologic region. In addition, the development of methodologies for quantification, measurement and hydrologic forecasting adapted to this kind of regions is also a challenging task.

INTRODUCTION

A good deal of the Earth's population inhabits large flatlands that very often supply huge quantities of food to humankind. Knowledge of the hydrologic processes in flatlands is of utmost significance for the socioeconomic development of these regions, in order to preserve their environmental characteristics and keep the sustainability of life.

Today, human activities have become one of the most negative influential factors for global changes in the biosphere. The overall results are land degradation and the overexploitation of fresh water resources, with the consequent worsening fresh water quality. All these harmful consequences are responsible for the great complexity of the hydrologic problems in extended flatlands. The same anthropic activities also affect the behavior of ecosystems and their capacity for supplying the inhabitants of an extended flatland with goods and services upon which the population depends. The need therefore exists for having interdisciplinary management tools for a better understanding of water resources, regarding these resources as abiotic and also as a service provided by ecosystems.

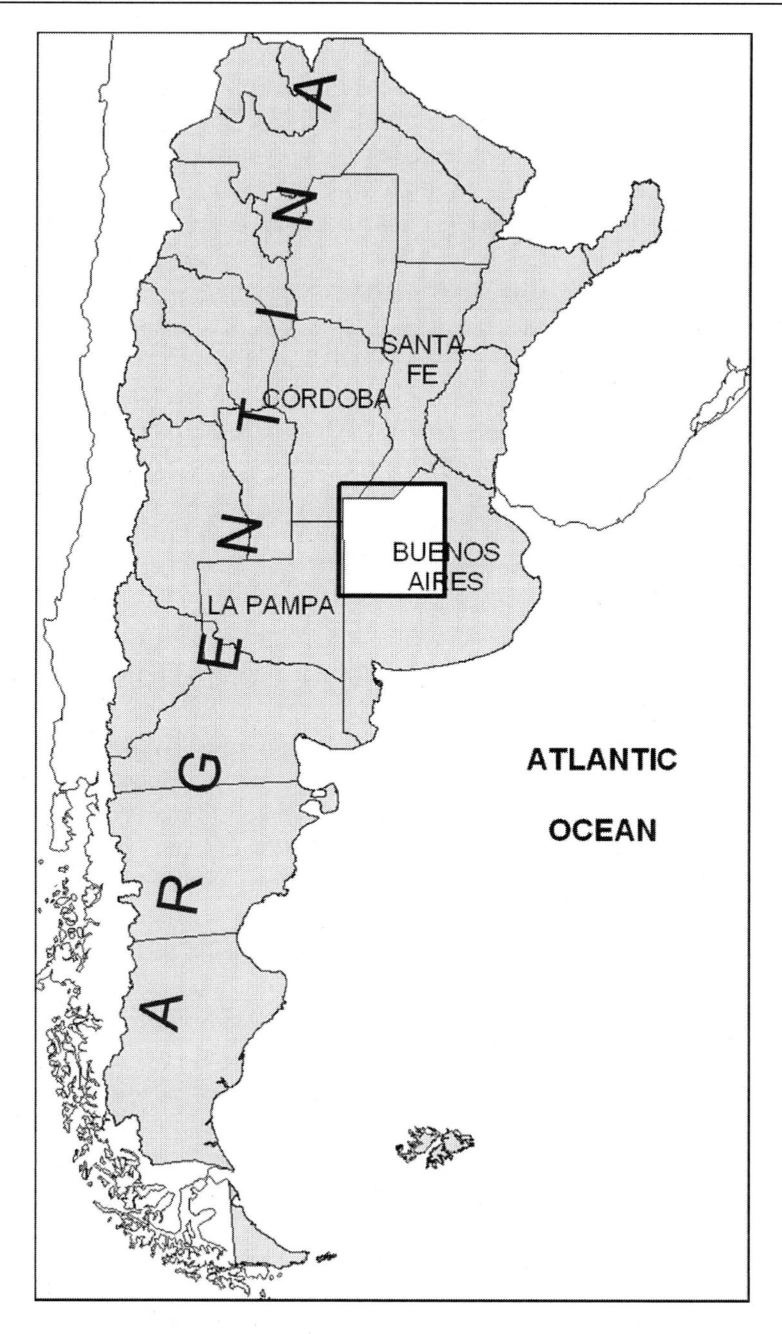

Figure 1. Location map of the sandy Pampa in Argentina.

The hydrology of flatlands presents particular features mainly associated with the low topographic slopes of these environments that must be taken into account in any evaluation process. A challenge therefore rises for the development of theoretical and applied hydrologic research into flatlands with the aim of suitable tools for dealing with all the complexity that water problems have in these ecosystems. Recent advances in flatlands hydrology aim to include groundwater as an important matter in environmental issues, and also to quantify infiltration processes, evapotranspiration and water transport in both the saturated and unsaturated zones (Kruse and Zimmermann, 2002).

The objective of this chapter is to put forward the challenge involved in hydrologic studies of flatlands, with a special emphasis on the difficulty of applying the traditional concept of a basin used for regions with a more abrupt relief. A sector of the Pampas Plain, locally known as "Pampa Arenosa" (Sandy Pampa), in the northwest of the Buenos Aires Province, Argentina (Fig. 1) is presented as a case study.

General Hydrological Conditions

Classical hydrology developed in Europe, where flatlands with low topographic gradients are scarce. Because of this, the principles of traditional hydrology are based on the response of surface runoff to rainfall. Afterwards, these principles were adapted and developed in North America, where they still remained valid because the natural conditions are similar to those of Europe. Therefore, classical hydrology tends to treat water flux processes in well defined drainage basins, partially composed of solid rock substrata with high slope lands. At the other end, it is possible to find large flatlands with very low topographic slopes developed on soft rock lands where surface runoff is generally poor. Whatever the hydrologic conditions, the definition of a flatland implies a morphological feature. A flatland is a plain relief of low topographic gradient. It is not usual to put a limiting value for this topographic gradient, but slopes less than 1% are traditionally indicated. It should be emphasized, however, that there exist flatlands with slopes much less than 1%, such as those found in the Pampas Plain (Argentina), which may be regionally on the order 1‰ or even less than 0.5‰. Within this region it can be found micro environments with slopes locally higher than those mentioned above.

In a mountain zone the outcrops of the geological units are easy to visualize in the field, which allows both their spatial position and extrapolation to the subsoil to be determined clearly. The situation is more difficult in a flatland because only a vegetated soil developed over recent sediments can be observed, and the geological units are not easily identified.

The main distinctive feature of the hydrology of flatlands is the low morphologic energy of the land, which makes the hydric energy dissipate through vertical, rather than horizontal, movements of water. Because of this, wetlands become temporary flooded with accumulation of salts (especially of sodium) near the surface and, in many cases, the development of swamps and shallow ponds. Surface water and groundwater should thus be considered as a unit. On the contrary, in arid zones the surface hydrology can be uncoordinated or even unconnected with groundwater, which determines the type and distribution of vegetation under natural conditions. Vertical movements in a flatland, such as evaporation, transpiration, infiltration and water transport between the unsaturated zone and the water table, prevail over horizontal superficial and groundwater movements. Flatlands are thus highly sensitive to climatic fluctuations (water deficit and surplus) as well as to anthropic activities.

All these features have led to make a distinction between the behavior of a hydrologic system in a mountainous area (a typical hydrologic system) and that in a flatland (a non-typical hydrologic system) (Fertonani and Prendes, 1983). The foremost reason for such a distinction is the topographic slope of the environment. In the hydrologic system of a mountainous area it is possible to recognize a significant slope that makes a morphogenetic energy capable of generating a drainage network, as well as surface runoff and erosive fluvial processes. On the contrary, in the hydrologic system of a flatland, the topographic slope is very low or nearly zero, which is insufficient for generating drainage, runoff and erosive processes. Therefore, the hydrology of flatlands shows features and hydrologic processes very different from those of the classical hydrology of a mountainous area. Consequently, the methods of quantification, measurement and forecasting require proper adaptations, making them different from the traditional methods. The scientific treatment of the hydrology of flatlands is complex and is still under development. Today's work on this subject should tend to promote a rational management of water, soil and vegetation for satisfying the needs of local inhabitants, preserve the environment and reach a sustainable development of the region.

Study Area

Regional Setting

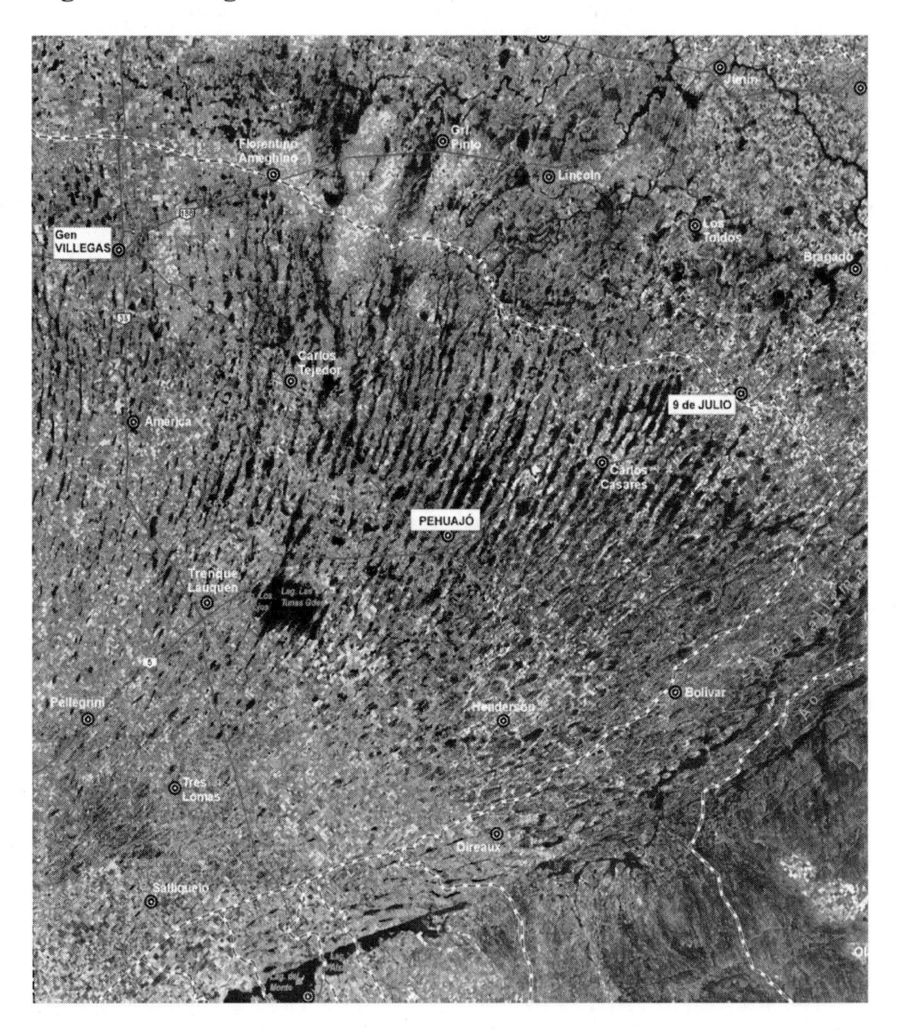

Figure 2. Satellite image of the sandy Pampa showing three towns: 9 de Julio, Pehuajó, and General Villegas.

The case study corresponds to the northwest region of the Province of Buenos Aires (Argentina) (Fig. 1). It is a flat environment that extends for about 50,000 km^2, located between 120 and 130 m over mean sea level, with a low topographic slope of about 0.1 m/km, and known under the traditional name of Pampa Arenosa (Sandy Pampa). Some minor topographic forms that are hydrologically important can be differentiated within the general morphology. There are slightly noticeable elevations alternating with depressions and "cañadas" (gullies) that give a relief of soft, somewhat aligned crinkles. They represent typical forms of paleodune environments. The region shows elongated ridges of ten kilometers long and two kilometers wide, with a prevailing southward direction (Fig. 2). The present wet climatic conditions contrast sharply with the origin of an eolic landscape associated with a medium Holocene climate of scarce available water (Zárate and Rabassa, 2005). The predominant features are the absence of a drainage network and the accumulation of water in blowouts or in inter dune openings.

From a hydrogeological point of view the region is covered with recent permeable sediments, fine sand and silts, forming dunes where several overlying eolic cycles have been recognized (Kruse et al., 2001). Finer sediment with a relative low permeability has been found in inter dune openings. The entire region is an arreic environment where floods occur in low lands during humid periods (Kruse et al., 2006) (Fig. 2). Sandy eolic sedimentary layers with a thickness ranging from a few centimeters to six meters are found at the surface. They lie over silts with variable proportions of sand, clay and usually calcareous materials, known as "sedimentos pampeanos" (Pampean sediments) (Fidalgo et al., 1975). The mean annual rainfall and temperature in the region are about 850 mm and 16 °C. The regional rainfall pattern shows alternating dry and humid periods. However, in overall, the first half of the twentieth century was drier than the second half.

The Problem of Floods and Droughts

Since the early 70's a wet climatic cycle characterized by an increase in rainfall has been recognized in the Pampean region. An analysis of the mean annual rainfall records between the periods 1941-1970 and 1971-2010 allows this wet cycle to be identified clearly. The increase in rainfall exceeds 150 mm/y.

A particular analysis of what happened in some regional towns illustrates the increase in rainfall. Figures 3, 4 and 5 show the annual series of precipitation in three towns: 9 de Julio, Pehuajó and General Villegas, respectively. The increasing trend in rainfall from the decade 1970-1979 is easily recognized. As a result, the region exhibits a clearly constant tendency to utilize these areas for agricultural activities. These areas have undergone a change in their hydrologic regime that has benefited the productive possibilities. As a counterpart, the increase in rainfall has produced floods in many local areas at different degrees and frequencies. Paradoxically, however, the floods are responsible for the decrease in lands available for agricultural activities, and are the cause of severe damage to roads and urban infrastructure.

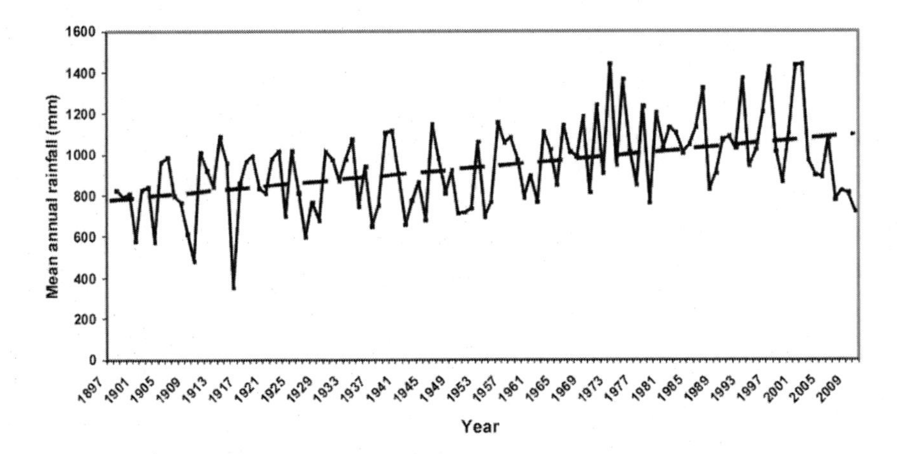

Figure 3. Mean annual rainfall for 9 de Julio.

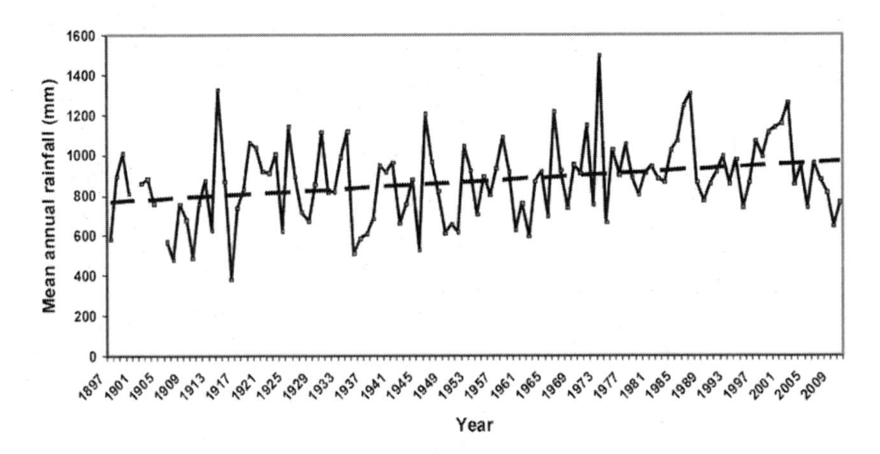

Figure 4. Mean annual rainfall for Pehuajó.

Hydrologic budgets are performed to estimate water surplus that will feed soil wetness and infiltration, or will be stored in low-elevation areas of the surface. In this regard, the reserves of available groundwater are a clear indication of hydric variations, and as such they have a direct influence upon agricultural activities. Changes in the water table are a direct consequence of changes in rainfall and water surplus. During a wet period the rise of the water table can generate floods in depressed zones (Fig. 2), which hinders the use of these sectors for agricultural activities; but at the same time, the rise of the water table allows high wetness values to be retained in high land zones, which permits the need for water in cultivation areas to be satisfied suitably. The inverse is true during a dry period. The decrease in rainfall and water surplus during a dry period produces the withdrawal or the lost of water from the depressions and the subsequent fall of the water table, which makes groundwater seepage volumes decrease. Instead, the rise and outcrop of the water table during a wet period with large water surplus increase infiltration and make water to stay in low-lying sectors. Even in the sectors with an apparently dry surface, the water table is close to the surface (between one and two meters deep), which is a clear indication of a flooding hazard. Under these circumstances, water occupies the whole area of natural ponds, exceeding in many opportunities their sizes and flowing into neighboring ponds. During a dry period, instead, evaporation exceeds water inflow, the water table falls and the flooded areas become reduced. In this case, the threat of flooding decreases.

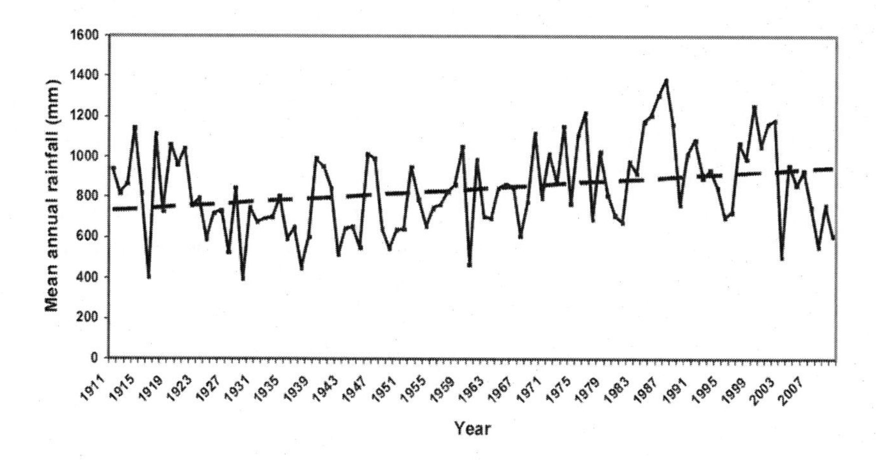

Figure 5. Mean annual rainfall for General Villegas.

Hydrologic Behavior

The study area, which supports a very important productive system, is an extended flatland periodically subject to strong hydric anomalies. The landscape is characterized by low topographic slopes, the absence of streams and the presence of permeable materials on the surface. Because of this, water from rainfall does not tend to flow along the surface but it moves vertically as infiltration, evaporation and transpiration. Besides, water accumulates in low-lying zones (depressions and ponds) and in the subsoil (changes of the water table level). It should be noted that as a result of these features and the presence of a shallow water table, surface water and groundwater are closely related and should be treated as a whole.

The hydrological situation of this region has been aggravated by the behavior of a network of canals originally constructed to drain the excess of water from heavy rainfall periods. Not only have the results been far from satisfactory, but also they have been harmful to the region. The real consequences of this rather old idea of constructing canals for drainage purposes is that during heavy rainfall periods, these canals do not convey the excess of water to the sectors originally chosen for receiving this surplus, but to the sectors that are already under a saturation state. Therefore, apart from climatic factors and the geomorphologic features, the hydrologic state of the region is strongly affected by human activities. Also, roads, canals or railways have modified the natural distribution of surface water with disastrous consequences. It should be taken into account that this flatland is not homogeneous, but has different topographic, morphologic and hydrogeological features, as well as variations in the distribution of ponds and natural drainage networks. Climate fluctuations between wet and dry periods are critical for agricultural activities because the amount and quality of water determine the eventual use of the significant technological tools installed for the enhancement of the regional production system (Forte Lay et al., 2008). Hydrologic budgets should be introduced as a fundamental step to understand the overall regional behavior. According to the physical characteristics of the area under analysis, the hydrologic budget is defined in the following terms:

Inflow – Outflow = + Change in storage capacity

The main variables of the budget are:

a) Inflow: precipitation
b) Outflow: evapotranspiration, stream runoff, groundwater flow
c) Storage capacity:

1. Groundwater storage: representing the volume of water likely to be stored between the water table and the land surface (unsaturated zone). This is a highly relevant variable because of its continuity all over the regional land and because of the porosity of the surface medium.
2. Superficial storage: representing the volume of water that can be potentially stored in ponds, "cañadas" (gullies) and low-lying zones that store surface runoff temporarily, slowing down infiltration and groundwater flux.

It is possible to consider the overall behavior of the region under wet and dry hydrologic periods. Under a wet period, inflow exceeds outflow. The degree of the difference generates a decrease in the groundwater storage capacity (rise of the water table) and the surface storage capacity (increase of flooded areas).

Under a dry period, outflow exceeds inflow. The deficit is covered by the geological water reserves, which makes groundwater and surface storage capacities increase, by the fall of water table and by the decrease of pond's sizes.

In order to know the present hydrologic situation it is of the utmost importance to determine the variations of rainfall, the influence of evapotranspiration and the evolution of the water table and the areas covered by surface water. It is thus essential to determine the water budget and the vertical exchange of water.

To do this, variations of the water table level and of the water storage depressions extent should be measured in addition to conventional observations. In this context, satellite images are an important tool for measuring changes in water storage depressions. The example of the Sandy Pampa region shows the importance of monitoring hydrologic stages by multitemporal analysis.

CONCLUSION

The case study presented above illustrates the importance of vertical water movements (evaporation, transpiration, infiltration, exchanges between the water table and the unsaturated zone) and variations in superficial and groundwater storage in a flatland environment.

From the point of view of hydrology, the most noticeable feature of the Sandy Pampa is the existence of alternate dry and wet periods. During dry periods the deepening of the water table makes agricultural demands difficult to be satisfied.

During wet periods, instead, the water table is shallower, usually producing overflows and affecting agricultural production negatively. Because the water table is often very close to the surface, groundwater and surface waters (ponds) are strongly related, and so they should be treated as a single unit. Satellite images can be used to track fluctuations in the ponds extent. These fluctuations are an important sign of hydrologic behavior.

Due to the low topographic gradients of this flatland environment, about 0.1 m/km, there is no enough water energy to generate neither an integrated drainage network nor an apparent surface runoff. Because of this, it is not possible to define the study region in terms of the traditional concept of a hydrologic basin. Instead, it seems more suitable to define a water budget for a given sector within the region.

The most significant terms in such a water budget would be precipitation, evapotranspiration, infiltration and variations in superficial and groundwater storage. Therefore, the hydrologic studies of a flatland should aim at measuring these variables carefully. An adequate network for measuring the water table level and the extent of surface water depressions (ponds), meteorological data and the use of serial satellite images will allow the evolution of hydrologic phenomena to be analyzed and more reliable forecasts to be made. In this way, a rational and sustainable management of water resources in flatlands can be achieved.

References

Fertonani, M; Prendes, H. 1983. *Hidrología en áreas de llanura. Aspectos conceptuales teóricos y metodológicos.* Coloquio Internacional sobre Hidrología de Grandes Llanuras. Olavarría, Argentina, 1983, Tomo I: 120 – 156.

Fidalgo, F; De Francesco, O.; Pascual, R. *Geología superficial de la llanura bonaerense.* VI Congreso Geológico Argentino, 1975, Relatorio: 103 - 138.

Forte Lay, JA; Kruse, E; Aiello JL. *Geo Journal*, 2007, 70(4), 263-271.

Kruse, E; Forte Lay, JA; Aiello, JL; Basualdo, A; Heinzenknecht, G. *Hydrolog. Sci. J.*, 2001, 267, 531 – 536.

Kruse, E; Zimmermann, E. *Hidrogeología de Grandes Llanuras. Particularidades en la Llanura Pampeana (Argentina).* Workshop. Groundwater and Human Development. Mar del Plata (Argentina), Proceedings of the XXXII IAH Congress, 2002, 2025 - 2038.

Kruse, E; Forte Lay, JA; Aiello, JL. *Hydrolog. Sci. J.*, 2006, 302, 91 – 97.

Zárate, M; Rabassa, J. *Geomorfología de la Provincia de Buenos Aires.* XVI Congreso Geológico Argentino, 2005, Relatorio: 119-138.

In: Basins
Editor: Jianwen Yang

ISBN: 978-1-63117-510-7

Chapter 6

HYDROCHEMICAL STUDY IN SURFACE WATERS OF BARREIRO, ARAXÁ CITY, MINAS GERAIS STATE, BRAZIL

Luis Henrique Mancini and Daniel Marcos Bonotto*
Departamento de Petrologia e Metalogenia,
Universidade Estadual Paulista (UNESP), Câmpus de Rio Claro,
Rio Claro, São Paulo, Brasil

ABSTRACT

The chemical evolution of the Earth´s surface and, consequently, of the hydrographic basins, is controlled by several factors, among which are the interaction of rainwaters, the atmosphere and the continental crust. Different climatic conditions affect the denudation, chemical weathering and physical erosion in hydrographic basins. The present and future design of the hydrographic basins is affected by chemical weathering processes and such information has been increasingly used by decision makers engaged in the management of our river basin systems. A lot of factors have been applied in models that utilize the concentrations of constituents like sodium, calcium, potassium, magnesium and total dissolved solids for estimating the chemical weathering fluxes taking place in the drainage basins. On the other hand, the population growth in all countries has caused an increased concern on the quantity and quality

* Corresponding Author: Email: danielbonotto@yahoo.com.br.

of the water resources in hydrographic basins. Well-defined anthropogenic inputs and/or pollution from diffuse source loads have commonly affected the chemistry of water bodies, requiring special attention by people involved in the river basin management. This chapter reports the results of a hydrochemical study held at the Barreiro hydrographic basin, Araxá city, Minas Gerais State, Brazil. The surface water quality in the site is not only affected by geogenic inputs but also by anthropogenic sources that are mainly coupled to activities involving the mining and production of fertilizers, which take place in alkaline rocks belonging to the geological formations of the region.

INTRODUCTION

The hydrosphere can be viewed as a system of different reservoirs in which water, solutes and energy are exchanged by the hydrological cycle. The total amount of water in the hydrosphere is estimated at 1.4×10^9 km^3, of which 96% resides in the oceans [1]. The remaining 4% consists of freshwater, which exists and moves only by the virtue of the continuous distillation process that turns salt water into a freshwater by evaporation and subsequent condensation [1]. The estimated freshwater volume in the hydrosphere is 40×10^6 km^3, most of which (~69%) is locked up in ice caps, icebergs and glaciers [1]. All river waters in world comprise only about 0.005% of the total amount of freshwater [1]. Such global scenario appears to be incompatible with the size and water volume discharged by most of the well-known world rivers.

For instance, let us consider the Amazon River. It starts flowing in the Andes Cordillera in Peru (5,600 m of altitude), intercepts a short part of south Colombia and flows through the northern portion of Brazil in the west-east direction until its discharge in the Atlantic Ocean between Amapá and Pará States. It receives the following names, along its 6,840 km of extension: Lloqueta, Apurimac, Ene, Tambo, Ucayali, Solimões and Amazonas. It is the largest in the world either in extension or amount of water discharged (~200,000 m^3/s during the rainy season) [2]. The total number of tributaries corresponds to about 7,000 along its flow, where its discharge is ~60 times higher than that of the Nile River, the second largest in the world. The width of the Amazon River is variable, ranging from 13 to 50 km in Manaus city (Amazonas State, Brazil), depending on the dry and rainy season. It is narrower (1,800 m) and deeper (50 m) at Óbidos city (Pará State in Brazil), where the discharge is 200,000 m^3/s. The average water level high is 10 m, reaching 16 m in the rainy season. The amount of sediments released into Atlantic Ocean is huge, circa 800 billion tons per year [3].

As a consequence of the development of the chemical industry, several constituents were released to water bodies and many countries established standards focusing on the drinking water quality. However, after the Second World War, the WHO (World Health Organization) published *International Standards for Drinking-water* in 1958, 1963 and 1971. In 1982, WHO shifted its focus from 'International Standards' to 'Guidelines', and the main reason for that was the advantage provided by the use of a risk-benefit approach (quantitative or qualitative) to the establishment of national standards and regulations. Specifically, the application of the Guidelines to different countries should take account of the sociocultural, environmental and economic circumstances pertinent to those countries. Thus, it was in 1984-1985 that WHO published in three volumes the first edition of *Guidelines for drinking-water quality* with the aim to supersede earlier European and international standards. Volume 1 contained guideline values for various constituents of drinking water, Volume 2 presented the criteria monographs prepared for each substance or contaminant on which the guideline values were based, and Volume 3 was concerned with the monitoring of drinking water quality in small communities, particularly those in rural areas. The last edition of the WHO *Guidelines for drinking-water quality* occurred in 2008. It again expressed the global concern on the water quality in the chemical perspective, with implications for the drainage basins utilized in water-supply systems.

Araxa City and Barreiro Area

Araxá is a city located in southeastern Brazil at 19°35'33"S and 46°56'26"W, possessing an altitude of 973 m (Figure 1). It has a mild weather throughout the year. It is 375 km distant from Belo Horizonte (W), 649 km from Brasília (SE), the country´s capital, 830 km from Rio de Janeiro (NW) and 581 km from São Paulo (NW). The town was named after the Indian tribe "Araxas" who lived there at the time it was first discovered. Its meaning is "the place from where the sun is seen first".

For many years, the place was inhabited by the Indian tribe "Araxas". In the 18th century, the cattle breeders and farmers arrived, attracted by the quality of the fertile soil, the climate and the mineral waters. In 1791 the "Freguesia de São Domingos do Araxá" was created, in 1831 the place became a "village", and in 1865 the village was upgraded to the town category. In 1944, the President Getúlio Vargas inaugurated the "Grand Hotel Hydrothermal Complex" in a place known as Barreiro (Figure 1). In the 60's and 70's, the mining of niobium and phosphate fertilizer expanded and also the industrial activity.

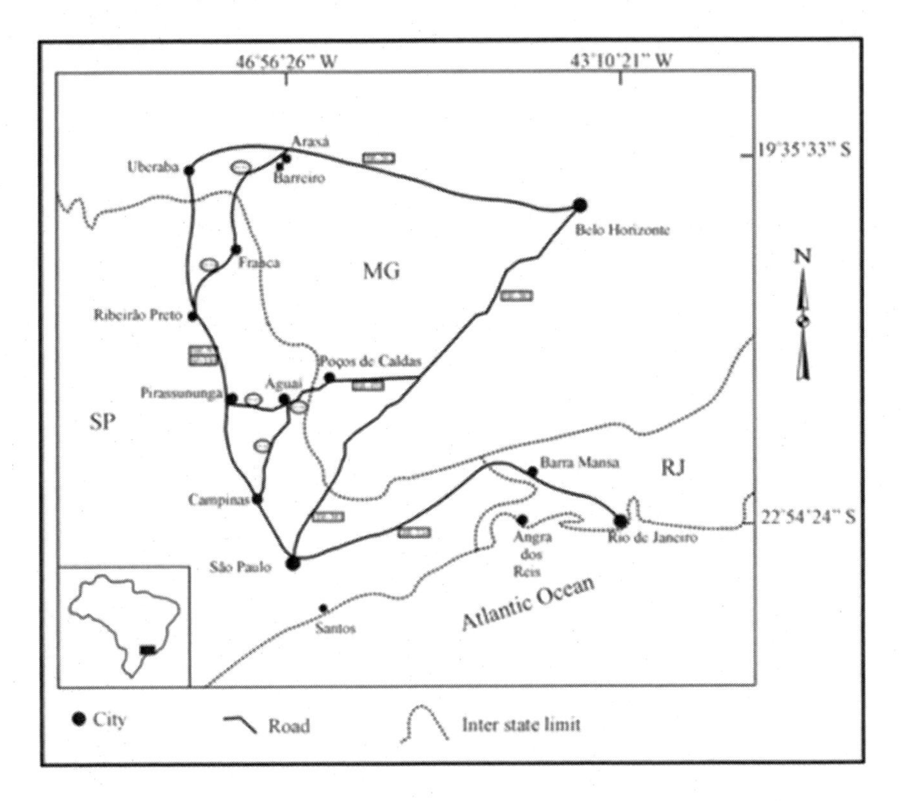

Figure 1. Location of Araxá city and Barreiro area, Minas Gerais State, Brazil. Modified from [4].

The Araxá region, which was brought into cultivation in the gold rush days by *bandeirantes* (the armed pioneers who pushed forward into the Brazilian hinterland), became a place of refuge for runaway slaves, who banded together in *quilombos* (fortified camps) and soon began to threaten the communication lines between São Paulo and the *bandeirantes*' settlements in Goiás.

Later, the town became known for the love affair between Dona Beja (Anna Jacintha de São José), the "witch of Araxá", and the justice of the peace of the neighboring town of Paracatu [5] (Dannemann, 2009). In the early years of the 19th century, Dona Beja became one of the most notable women in Minas Gerais history and also a leading figure in the upper ranges of local society [5]. The love history and other historic facts about her life were focused in the 89 chapters of *Dona Beija* novel that was exhibited in 1986 by the Brazilian Manchete TV [6]. Mention to her life has also been provided by the Dona Beja Museum that is located in Araxá city center.

The Barreiro area is situated about 8 km from Araxá city center, being well-known by the occurrence of the Dona Beija springs, whose waters are highly radioactive, also containing dissolved bicarbonates, calcium and magnesium. In addition to a hydrothermal complex, Barreiro area is also characterized by the presence of phosphate and niobium mining activities that have been developed, respectively, by Bunge (past Serrana) and CBMM (*Companhia Brasileira de Metalurgia e Mineração*) companies.

Geological and Mineralogical Aspects of Barreiro Area

Alkaline rocks are found in southern Brazil in association with the Ordovician-Cretaceous sedimentary Paraná basin and mainly distributed along its margins. They vary considerably in age and composition, and have been assembled into distinct geographic provinces, such as the Alto Paranaíba Igneous Province, representing one of their most important provinces, particularly due to the presence of economical deposits bearing niobium, phosphate, titanium and rare earth elements [7]. The Alto Paranaíba Province includes the renowned carbonatite intrusion of Araxá, covering approximately 16 km^2, which is in general related to a NW-trending linear structure that borders the São Francisco cratonic area and that is thought to be active since late Precambrian times [7].

The Araxá complex has been previously reported in literature as Barreiro. It consists of a circular intrusion, 4.5 km in diameter, with the central part mainly formed by a carbonatite predominantly beforsitic in composition [7]. A complex network of carbonatite as concentric and radial dykes is variable in dimension, and small veins ranging from few millimeters to several centimeters in thickness are also present intruding either alkaline or country rocks. Additional lithologies include mica-rich rocks, phoscorites and lamprophyres [7].

The Barreiro area is located in a fractured zone tending to a major NW direction. The alkaline complexes of Tapira and Salitre I and II in Minas Gerais State and Serra Negra and Catalão I and II in Goiás State are situated in the same fractures zone that is characterized by several parallel faults whose size reaches hundreds of kilometers in length [4]. All these bodies comprise the also named Alto Paranaíba Alkaline-Carbonatite Province that is embedded in the *metassedimentary sequence*s of Araxá, Canastra and Bambuí Groups. Schists and quartzites are the main rock types associated with Araxá Group that is linked to the Barreiro area.

The weathering mantle resulting from the alteration of the alkaline-carbonatite rocks has exhibited accentuated thicknesses, locally higher than 200 meters. Its evolution is related to several peneplained phases that affected the region. The alteration material is primarily enriched in P_2O_5, Nb_2O_5, TiO_2, BaOand REE_2O_3. The distribution of these oxides in the intrusion area is irregular, reflecting the complex lithology of the alkaline association [8]. The weathering processes promoted the pyrochlore enrichment in the central portion of the circular structure that originated one of the largest world niobium deposits. The average mineralogical composition and niobium reserves are shown in Tables 1 and 2.

Table 1. Average mineral composition of the Araxá niobium ore [9]

Mineral	Weight %	Mineral	Weight %
Bariumpyrochlore	4.6	Monazite	5.0
Limonite, goethite	35.0	Ilmenite	4.0
Barite	20.0	Quartz	5.0
Magnetite	16.0	Others	5.4
Gorceixite	5.0		

Table 2. Total reserves of the Araxá niobium ore [9]

Category	Amount (tons)	Nb_2O_5 (%)
Measured	131,612,000	2.50
Indicated	41,793,000	2.49
Inferred	288,349,000	2.50
Total	461,754,000	2.50

The weathering mantle also contains a large phosphate reserve situated at the northwestern portion of the complex and comprising an area of about 2.5

km^2. Most minerals have suffered lattice modifications when submitted to weathering processes. A weathered cover and an apatite-shaped material have been identified in the region [8].

Two distinct materials exempt of apatite (despite the presence of P_2O_5) comprise the weathered cover. The superficial cover corresponds to a dark reddish layer whose maximum thickness is about 20 meters in the thicker areas. It is composed of hydrated iron oxides and clay minerals, also containing disperse or concentrated limonite blocks followed by a material sterile in apatite resulting from the extreme weathering of the original constituents.

The material in low lands has a reduced (or absent) thickness, but can reach up to 60 meters in the high lands. Such portion of the weathered cover shows enrichment in Fe_2O_3, SiO_2, Al_2O_3, TiO_2, BaO, REE_2O_3, Nb_2O_5 and P_2O_5 [8].

Table 3. Chemical composition (in %) of the phosphate ore samples in Araxá [10]

Constituent	A-1	B-1A	B-1B	B-2A	B-2B	C-1A	C-1B	C-2A
P_2O_5	16.00	11.00	8.97	10.40	12.00	13.10	9.47	10.40
CaO	18.60	16.40	14.80	17.40	20.60	23.50	32.50	22.00
SiO_2	5.25	19.20	24.00	16.70	16.40	14.10	9.02	18.50
Al_2O_3	2.29	1.91	2.66	1.92	2.21	1.35	1.04	1.86
Fe_2O_3	28.40	26.60	23.80	21.70	14.20	18.80	14.60	18.00
MgO	1.68	10.80	11.50	12.60	10.80	9.31	6.93	9.08
BaO	5.88	0.20	0.35	0.80	3.38	0.96	1.52	0.29
TiO_2	4.60	4.65	5.43	4.84	3.84	2.97	1.93	3.78
Na_2O	1.29	1.08	0.91	1.23	1.09	1.39	0.91	1.05
K_2O	0.17	0.97	0.89	0.59	0.80	0.43	0.40	0.95
Sr	0.63	0.21	0.15	0.28	0.37	0.38	0.28	0.23
S	1.05	<0.10	<0.10	0.15	0.55	2.93	0.24	<0.10
CO_2	1.28	0.71	1.03	2.20	3.46	4.84	15.65	5.75

Table 4. Chemical composition (in %) of a REEs-enriched average sample of the Barreiro area [12]

Constituent	Value	Constituent	Value	Constituent	Value
SiO_2	17.40	MnO	0.60	Sm_2O_3	0.19
Al_2O_3	14.70	PbO	0.65	Pr_6O_{11}	0.49
P_2O5	9.93	SO_3	0.41	Y_2O_3	0.19
Nb_2O_5	2.21	ZrO_2	0.47	U_3O_8	0.052
Fe_2O_3	11.60	CaO	0.91	ThO_2	0.60
BaO	5.58	CeO_2	6.28	H_2O^+	9.87
SrO	2.21	La_2O_3	3.80	Total	96.662
TiO_2	6.94	Nd_2O_3	1.58		

A zone containing apatite is immediately found below the superficial weathered cover. It has been identified by the presence of significant CaO levels.

The mineralized material tends to be a dark brown color, whose P_2O_5 content above 20% implies more than 50% in apatite. Non-colored micaceous-shaped materials are found below tenths of meters, but they tend to modify the color to gold or clear brown down the profile. Another level indicating less accentuated weathering conditions has also been recognized downwards. The micaceous minerals vary from greenish brown to dark green color, reaching a 20-30 meters thickness. They are suddenly succeeded by a new greenish colored material containing carbonates, frequently in 5-m intervals. The carbonates represent a significant fraction of the minerals accompanying apatite. They are within the 10-30 m thick weathered mantle.

Following this level, a layer of semi-cohesive material occurs that is not included in the weathering mantle. It preserves the original structures, configuring an almost weathered 10-20 m thick rock level, from which fresh carbonate and *glimmeritic* rocks are found [8]. Table 3 reports the chemical composition of these materials.

The Rare Earth Elements (REEs) are abundant in three main sites of the Barreiro area: left margin of Mata stream (E-W direction); central portion of the complex close to the niobium deposit; and southeastern portion of the niobium deposit [11].

Table 4 shows the chemical composition determined by X-ray fluorescence of a REEs-enriched average sample of the Barreiro area.

RADIOACTIVITY AT BARREIRO AREA

The pyrochlore is a complex mineral group, whose formula may be expressed by $A_2B_2X_7$. Several changes may occur involving A and B, for instance, A(Na, Ca, Ba, K, U) and B(Nb, Ta, Ti). Barium-pyrochlore is abundant in Araxá, where Ba and Nb dominate the A and B positions, respectively [12]. Studies held by IPR (*Instituto de Pesquisas Radioativas*, Belo Horizonte, Minas Gerais State, Brazil) have indicated different pyrochlore generations responsible for the uranium presence. Thus, this radioelement can occur widely disseminated in the mineral matrix as a trace constituent (100 ppm U_3O_8) or may characterize an uraniferous pyrochlore with the U_3O_8 concentration reaching up to 2%.

The occurrence of secondary uranium minerals like autunite and uranocircite has been identified in some areas, where they are filling rock fractures mainly containing apatite and baritine. Structural controls have exerted an important role in the uranium mineralization that has been verified when the host rock is greatly faulted. Under such a circumstance, vadose waters originating from the uraniferous pyrochlore deposit are able to find percolation conditions for later deposition of phosphate uraniferous minerals. All U-mineralization is located above the water table at a 27-m maximum depth. Autunite dominates in these zones where the average U_3O_8 concentration is 0.11%. The mineral exhibits thin milimetric blades of pale green color preferentially filling the rock fractures and voids.

U_3O_8 contents varying between 150 and 200 ppm have also been reported in the large phosphate deposit at Barreiro area, which has been extensively explored by Bunge (past Serrana). Uranium also occurs associated with the REEs at the central-northern portion of Barreiro area. The genesis of this deposit is linked to the presence of the REEs-enriched carbonatite that formed a closed ellipsoidal structure. In terms of morphology, the deposit is characterized by the horizontal settlement of several enrichment levels, with alternating layers of enriched and depleted U-REEs levels. This sequence is sometimes interrupted by small accommodation faults originated by dissolution of the protore constituents. The most important movements have only affected the southwestern portion of the deposit, where the south border suffered about 100-m displacement due to the transcurrent faulting. The probable calcium carbonate dominance in the protore allied to the relative barium and iron oxides absence caused a notable REEs enrichment during the weathering processes.

Spring Waters at Barreiro Area

The inauguration of the "Grand Hotel Hydrothermal Complex" in 1944 at the Barreiro area was coupled to the existence of two major springs having different hydrochemical composition: Dona Beija and Andrade Júnior.

In fact, Dona Beija spring corresponds to the first mineral exploration mark at the Barreiro area. These exploited mineral waters are an important natural and historic patrimony with significant economic and touristic impacts to Araxá city.

The water table interception at the terrain surface due to the topography causes the spring seepage. The water has low salinity but high radioactivity [13] and the discharge is sustained by the recharging rainwater. Low residence time in the aquifer is expected for these spring waters, where 2-3 km is estimated as the maximum recharge-discharge distance. The aquifer system has been classified as granular, un- and semi-confined, mainly in the intrusive body domain (2 km-diameter, variable thickness, reaching up to 200 m at southern portion, and close to the niobium mine) [14].

The aquifer may be locally semi-confined by clayey levels of the weathered mantle. The recharge occurs due to direct infiltration in areas of higher altitude and the flow lines at subsurface converge to the Grand Hotel lake and Sal stream. One main recharge area is the 70 ha "green ring" site located upwards of Dona Beija spring, which allows a significant aquifer replenishment, also contributing the waters of the "E", "F" and "BCM" dams and of the phosphate mine dyke [14].

The Andrade Júnior spring is situated in front of the Grand Hotel, possessing a genesis distinct of that of Dona Beija spring. The waters are thermal (33 °C), alkaline and sulfured, exhibiting high salinity. The water circulation is deep, outcropping in an area where the rock has not suffered significant alteration.

According to [14], the aquifer is fractured, and un- to semi-confined, mainly occurring in rocks around the carbonatite complex. The recharge is due to direct infiltration of rain waters in the little rocks exposures, as well by infiltrating waters in the quartzite residual soil. The groundwater flow is towards several directions as the ore body is circular and exhibits different altitudes.

HYDROCHEMICAL SURVEY AT BARREIRO AREA

The drainage area for surface waters occurring at Barreiro area comprises two major portions (Figure 2): 1) an eastern one, consisting of Baritina and Mata streams and respective tributaries; 2) a western one, consisting of Cascatinha and Borges streams. Cascatinha, Borges and Baritina streams drain the mining and waste disposal sites of the phosphate mine. Baritina stream also receives waters from the niobium mining, which influences Mata stream too. All streams flow into the Grand Hotel direction, originating Sal stream that corresponds to the unique Barreiro area exit. There are four dams at Barreiro area: two at Baritina stream (eastwards) - BCM dam and phosphate mine dyke; two at Cascatinha stream (westwards) - "E" and "F" dams. Two lakes have been established for composing the area landscape: Grand Hotel lake formed by the waters of Dona Beija spring, and the lake formed by the waters of Andrade Júnior spring.

Two sampling campaigns of surface waters were realized in this study held at Barreiro area. The first was performed in the wet period of 16 to 19 February 2000, whereas the second in the dry season from 12 to 16 September 2000. Twelve samples were *in situ* analyzed for physical parameters and stored in 1-L polyethylene flasks for chemical analyses. The sampling points (Figure 2) were chosen from the drainage analysis and location of the dams in the streams, spreading from the high lands (low course) up to the Barreiro area exit at Sal stream. Table 5 describes the location of the sampling points for surface waters at Barreiro area.

The temperature and conductivity readings were performed *in situ* using YSI probe model 3000 T-L-C, whereas pH and redox potential Eh data were acquired by Digimed portable meter. The water samples for chemical analysis were transported up to LABIDRO-Isotopes and Hydrochemistry Laboratory, Rio Claro city, São Paulo State, Brazil. They were filtered in 0.45 μm Millipore membrane for retaining the suspended solids, whereas the following parameters were characterized in the liquid phase: major cations and anions, silica and tannin/lignin. The Hach DR 2000 spectrophotometer was utilized for yielding data for silica, tannin/lignin, alkalinity (as bicarbonate), magnesium, calcium, barium, nitrate, sulfate and phosphate. The methods adopted are described in [16].

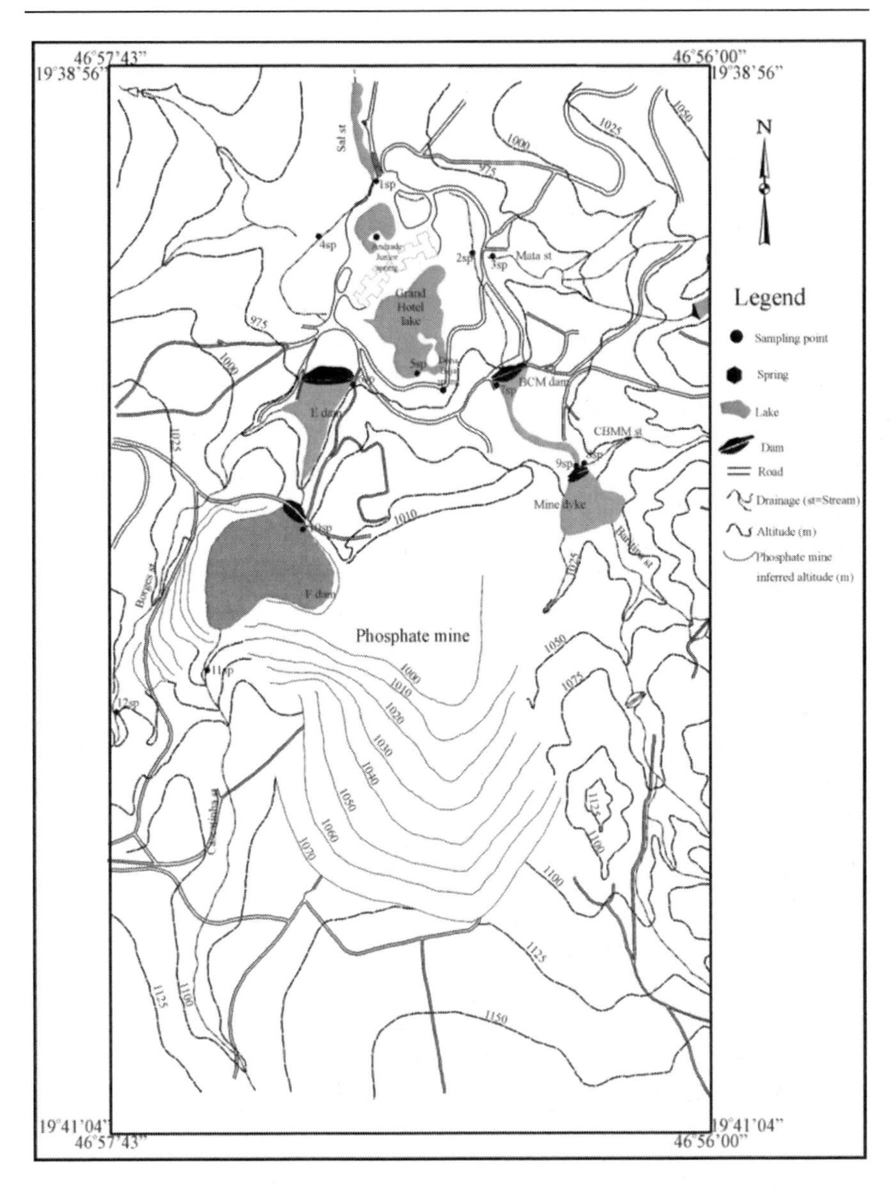

Figure 2. Location of the sampling points for surface waters at Barreiro area, Araxá city, Minas GeraisState, Brazil. Modified from [15].

The sodium measurements were done by potentiometry through the Orion Model 9811 combined electrode, coupled to Analion Model IA 601 meter [17]. The anions chloride and fluoride were also evaluated by potentiometry, following the technique described by [18].

Table 6 reports the results of the analyses of the physical parameters, whereas Tables 7 and 8 show the results of the chemical analysis.

Table 5. Location of the sampling points for surface waters at Barreiro area, Araxá city, Minas Gerais State, Brazil

Sample code	Latitude - Longitude	Description
1sp	19°38'39"S - 46°57'01"W	Salt stream, next to Barreiro church
2sp	19°38'44"S - 46°56'50"W	Baritina stream, downstream Mata stream, next Grand Hotel
3sp	19°38'46"S - 46°56'49"W	Mata stream, next Previdência Hotel
4sp	19°38'46"S - 46°57'04"W	Cascatinha stream, downstream "E" dam, next Barreiro grove
5sp	19°38'58"S - 46°56'56"W	Grand Hotel lake, behind Grand Hotel
6sp	19°38'59"S - 46°57'06"W	Barragem "E" dam, above Super Água water bottling company
7sp	19°38'58"S - 46°56'46"W	"BCM" dam, formed by past decantation dykes of Camig
8sp	19°39'00"S - 46°56'39"W	CBMM stream, close to the phosphate mining dyke
9sp	19°39'06"S - 46°56'39"W	Phosphate mine dyke exit, next past Camig installations
10sp	19°39'13"S - 46°57'10"W	Barragem "F" dam, Cascatinha forest
11sp	19°39'28"S - 46°57'21"W	Falls formed by Cascatinha stream at Cascatinha forest
12sp	19°39'32"S - 46°57'30"W	Borges stream, within Bunge (past Serrana) area

Table 6. Physical parameters in surface waters of Barreiro area, Araxá city, Minas Gerais State, Brazil

Sample code	Temperature (°C)	Conductivity (ms/cm)	pH	Eh (mV)
Sampling date: 16-19 February 2000				
1sp	25.0	0.348	7.33	164
2sp	23.2	1.500	6.60	192
3sp	25.1	0.511	7.15	164
4sp	26.8	0.336	7.32	158
5sp	29.9	nm	7.07	180
6sp	28.0	0.084	7.13	120
7sp	27.6	0.143	7.28	161
8sp	23.7	0.294	6.57	201
9sp	27.9	0.632	7.06	175
10sp	28.3	0.081	7.15	119
11sp	20.9	0.048	7.42	149
12sp	22.7	0.091	7.25	142
Sampling date: 12-16 Sep. 2000				
1sp	25.5	0.386	7.20	144
2sp	24.0	0.437	7.06	101
3sp	24.7	1.610	6.73	119
4sp	26.2	0.344	7.68	143
5sp	26.5	0.349	8.15	99
6sp	24.6	0.121	8.76	64
7sp	25.8	0.156	7.48	120
8sp	24.0	0.363	6.55	232
9sp	22.0	0.132	7.20	200
10sp	24.2	0.109	9.70	71
11sp	22.4	0.059	7.20	186
12sp	24.5	0.151	7.65	169

nm – not measured.

Table 7. Chemical parameters (mg/L) in surface waters of Barreiro area, Araxá city, Minas Gerais State, Brazil. Sampling date: 16-19 February 2000

	1sp	2sp	3sp	4sp	5sp	6sp	7sp	8sp	9sp	10sp	11sp	12sp
Na^+	7.50	31.62	15.40	0.65	0.49	6.04	0.49	1.54	0.45	0.27	0.24	0.60
K^+	10.50	30.00	38.00	nm	4.80	1.90	2.80	9.00	1.90	1.80	30.00	2.20
Ca^{2+}	0.03	0.03	<0.02	0.06	0.05	0.02	<0.02	0.06	<0.02	<0.02	<0.02	0.03
Mg^{2+}	0.39	0.03	0.05	0.25	<0.01	0.59	0.48	0.25	0.68	0.65	0.03	0.37
Ba^{2+}	3	7	6	6	5	2	5	6	6	3	5	3
HCO_3^-	60	24	46	76	68	36	68	40	44	52	20	34
Cl^-	31.9	>200	>200	36.4	44.4	0.1	3.8	85.7	2.3	0.1	0.1	0.1
F^-	0.20	0.14	0.15	0.22	0.20	0.24	0.18	0.16	0.20	0.18	0.09	0.13
NO_3^-	<0.1	0.4	0.3	0.3	<0.1	0.2	0.1	0.3	1.5	<0.1	0.4	0.3
SO_4^{2-}	7	41	19	2	7	8	2	7	12	6	41	5
PO_4^{3-}	0.30	0.08	0.26	0.30	0.20	0.40	0.36	0.66	0.47	0.25	0.08	0.03
SiO_2	16.2	14.8	13.8	17.4	16.0	10.5	12.8	12.4	13.5	10.3	7.0	11.9
TL^*	<0.05	0.40	<0.05	<0.05	<0.05	<0.05	<0.05	<0.05	<0.05	<0.05	0.10	0.20

*TL = tannin and lignin.

Table 8. Chemical parameters (mg/L) in surface waters of Barreiro area, Araxá city, Minas Gerais State, Brazil. Sampling date: 12-16 September 2000

	1sp	2sp	3sp	4sp	5sp	6sp	7sp	8sp	9sp	10sp	11sp	12sp
Na^{+}	1.90	7.50	86.60	1.10	1.80	0.30	10.70	2.20	0.20	0.30	0.30	0.90
K^{+}	3.10	5.40	3.62	4.28	4.37	1.80	1.76	2.87	15.30	1.52	4.10	4.21
Ca^{2+}	0.14	0.02	<0.02	0.02	<0.02	<0.02	<0.02	<0.02	<0.02	0.02	<0.02	<0.02
Mg^{2+}	<0.01	<0.01	<0.01	<0.01	0.02	<0.01	<0.01	<0.01	<0.01	<0.01	0.06	<0.01
Ba^{2+}	2	3	1	2	3	2	3	9	1	1	< 1	1
HCO_3^{-}	74	28	14	56	60	50	40	30	30	44	16	46
Cl^{-}	59.2	99.7	456.3	47.7	54.3	1.5	13.5	83.8	12.9	1.3	3.8	2.6
F^{-}	0.4	0.2	0.1	0.3	0.2	0.3	0.2	0.2	0.2	0.3	0.1	0.3
NO_3^{-}	0.2	0.1	<0.1	0.1	0.2	0.5	<0.1	0.3	0.1	1.0	0.1	0.1
SO_4^{2-}	16	22	38	1	1	3	9	4	4	2	1	3
PO_4^{3-}	>2.75	>2.75	>2.75	0.33	0.45	0.52	0.60	>2.75	0.53	0.52	0.21	0.36
SiO_2	14.5	12.9	12.9	12.4	14.0	6.0	7.5	10.8	6.1	8.4	6.0	12.5
TL^{*}	<0.05	<0.05	14	<0.05	<0.05	<0.05	<0.05	<0.05	<0.05	<0.05	<0.05	<0.05

*TL = tannin and lignin.

Physical Parameters Evaluation

Table 6 shows the temperature, conductivity, pH and Eh data measured for surface waters in the two field campaigns. The pH and Eh in the first sampling campaign were measured in laboratory due to the equipment unavailability for field readings during that time.

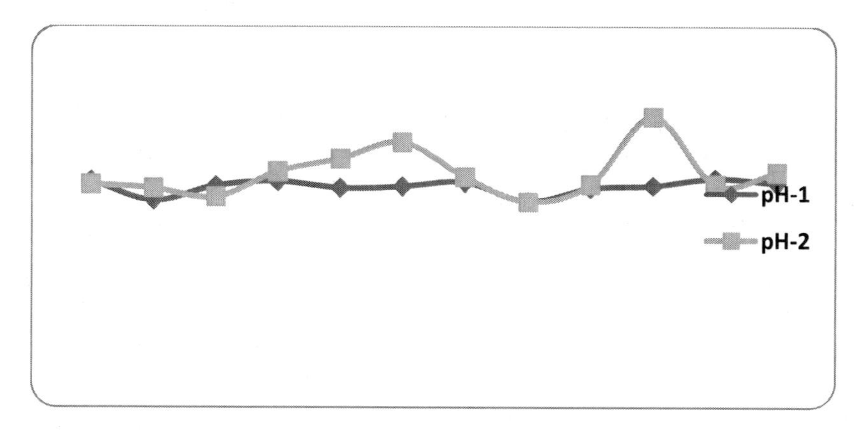

Figure 3. The pH data in surface waters at Barreiro area, Araxá, Minas Gerais State, Brazil. pH-1 (laboratory); pH-2 (field).

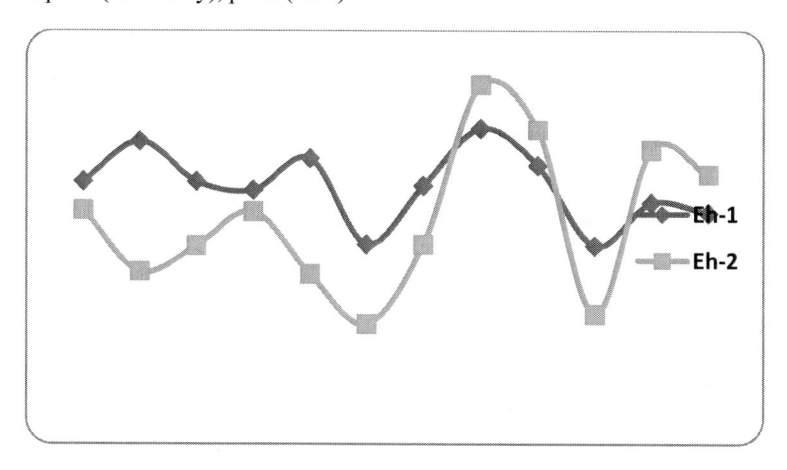

Figure 4. The Eh data (in mV) in surface waters at Barreiro area, Araxá, Minas Gerais State, Brazil. Eh-1 (laboratory); Eh-2 (field).

Figure 3 shows the pH data measured in field and laboratory. The values do not vary significantly for most analyses, suggesting the non-existence of differences among them. The higher amplitude variation may be attributed to seasonal effects. Figure 4 shows the Eh data measured in field and laboratory. The higher amplitude variation in pH found at sampling points 5sp, 6sp and 10sp was also identified in the Eh values.

Chemical Data Evaluation

Tables 7 and 8 show the results of the chemical analyses performed in laboratory for major dissolved cations and anions in surface waters. The hydrochemical data can be evaluated considering the two major drainage basins at Barreiro area, i.e., the eastern (Baritina and Mata streams plus respective tributaries) and the western (Cascatinha and Borges streams). Thus, two samples groups ae considered: Group 1 (eastern), consisting of monitoring points 2sp, 3sp, 7sp, 8sp and 9sp; Group 2 (western), consisting of monitoring points 4sp, 6sp, 10sp, 11sp and 12sp. The sampling points 1sp and 5sp have not been included in any groups as 1sp is a mixture of waters from groups 1 and 2, whereas 5sp is from the small lake formed by discharging waters of Dona Beija spring. Figures 5-10 display the chemical data corresponding to different groups and seasons (dry and wet periods) of the sampling campaigns.

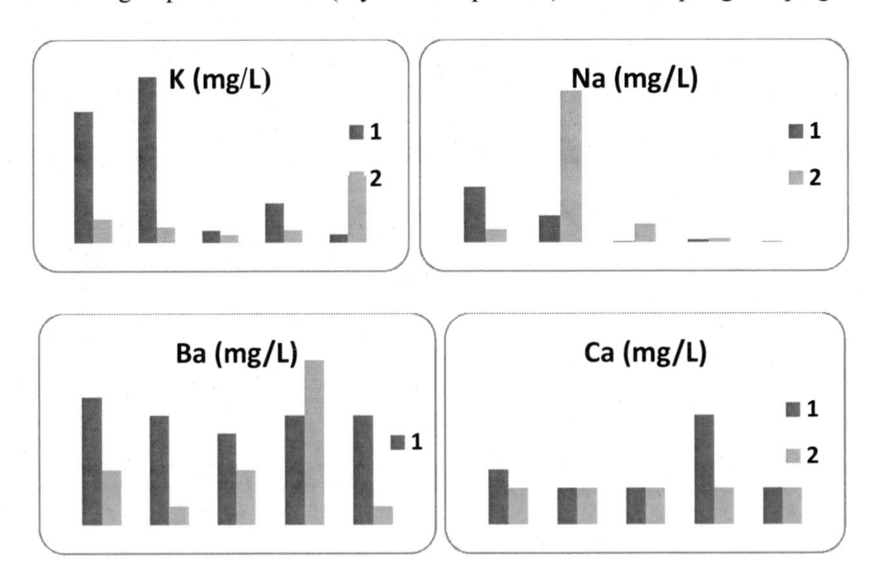

Figure 5. Potassium, sodium, barium and calcium contents in Group 1 waters of Barreiro area. 1 - February 2000; 2 - September 2000.

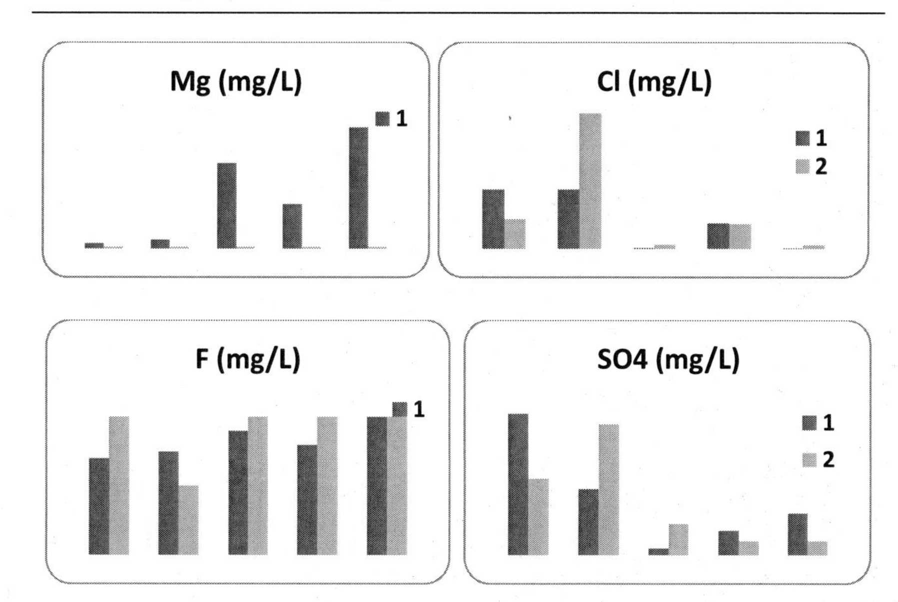

Figure 6. Magnesium, chloride, fluoride and sulfate contents in Group 1 waters of Barreiro area. 1 - February 2000; 2 - September 2000.

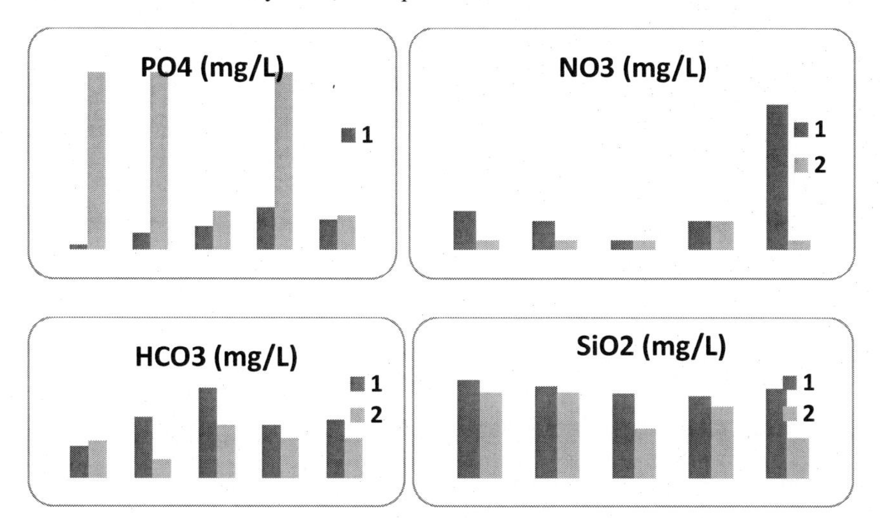

Figure 7. Phosphate, nitrate, alkalinity and silica contents in Group 1 waters of Barreiro area. 1 - February 2000; 2 - September 2000.

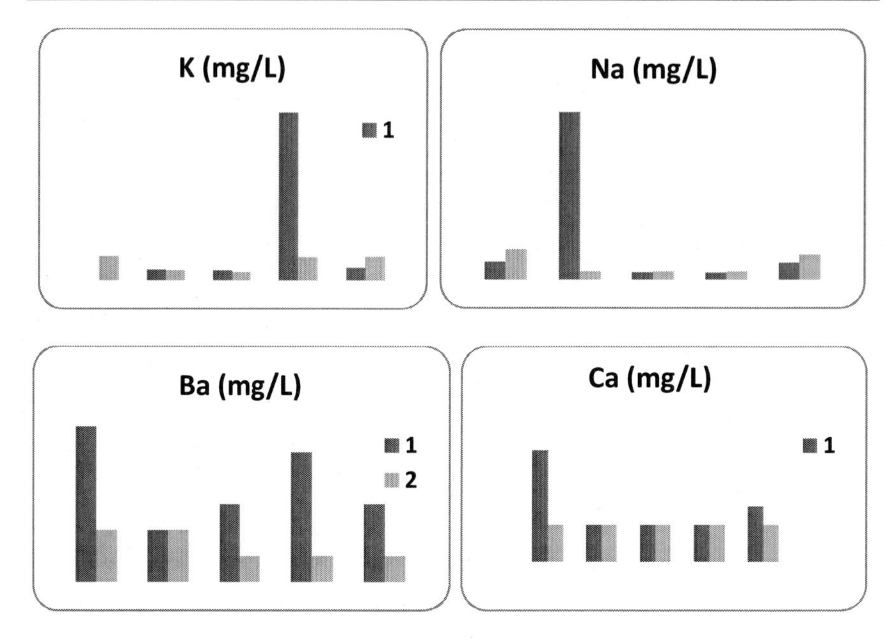

Figure 8. Potassium, sodium, barium and calcium contents in Group 2 waters of Barreiro area. 1 - February 2000; 2 - September 2000.

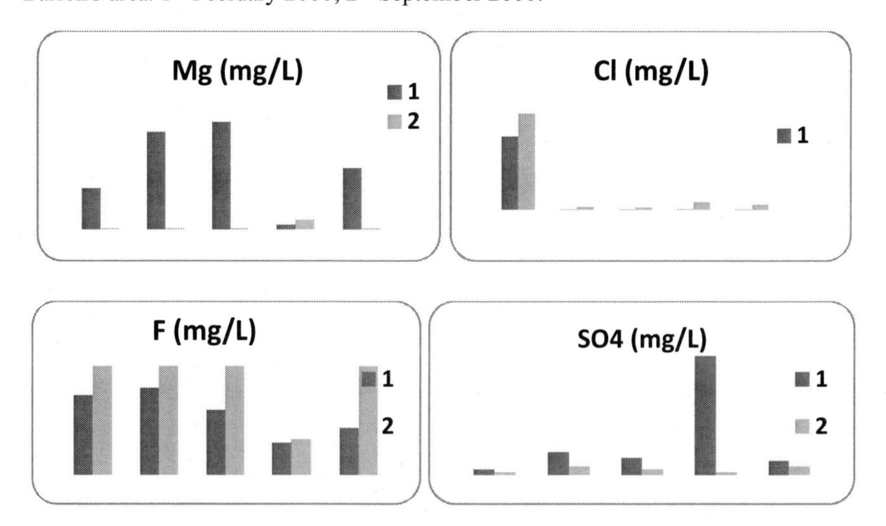

Figure 9. Magnesium, chloride, fluoride and sulfate contents in Group 2 waters of Barreiro area. 1 - February 2000; 2 - September 2000.

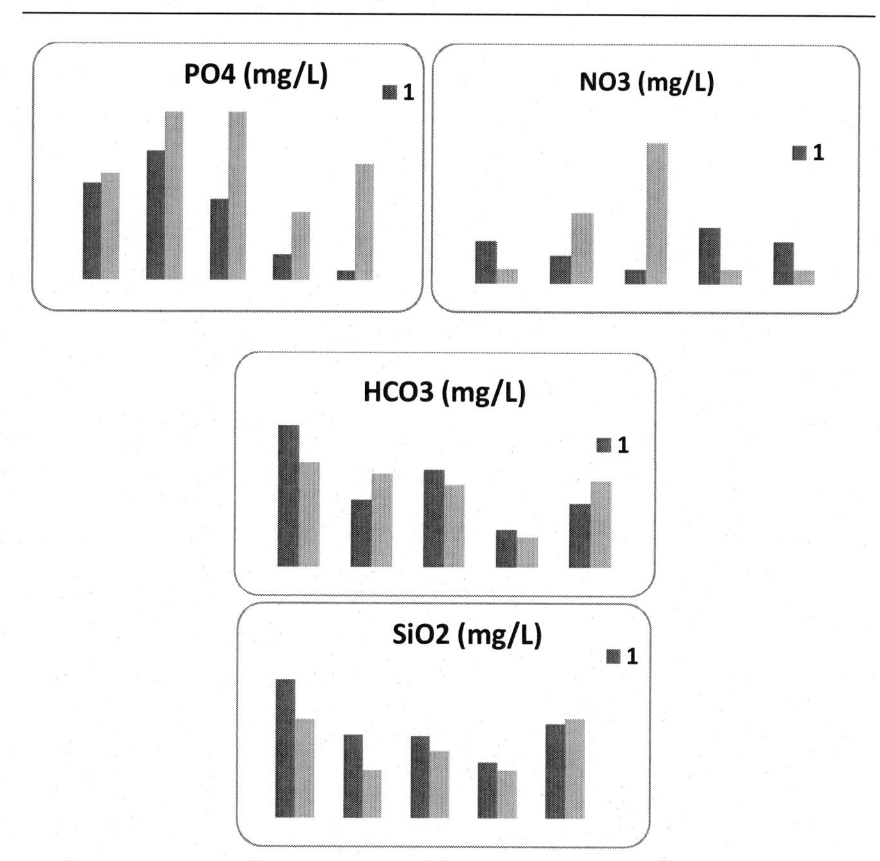

Figure 10. Phosphate, nitrate, alkalinity and silica contents in Group 2 waters of Barreiro area. 1 - February 2000; 2 - September 2000.

High Ba contents have been found in Groups 1 and 2, mainly in the wet period (Figures 5 and 8). This is an expected finding as the region contains several minerals hosting this element. In terms of some high Cl and Na contents, the values can be attributed to anthropogenic inputs as the drainages receive domestic wastes and mining-industrial effluents generated in Barreiro area (Figures 5, 6, 8 and 9). These waters also contain PO_4 due to the presence of minerals hosting P in the area; the values are higher in Group 1 waters during the dry period (Figures 7 and 10).

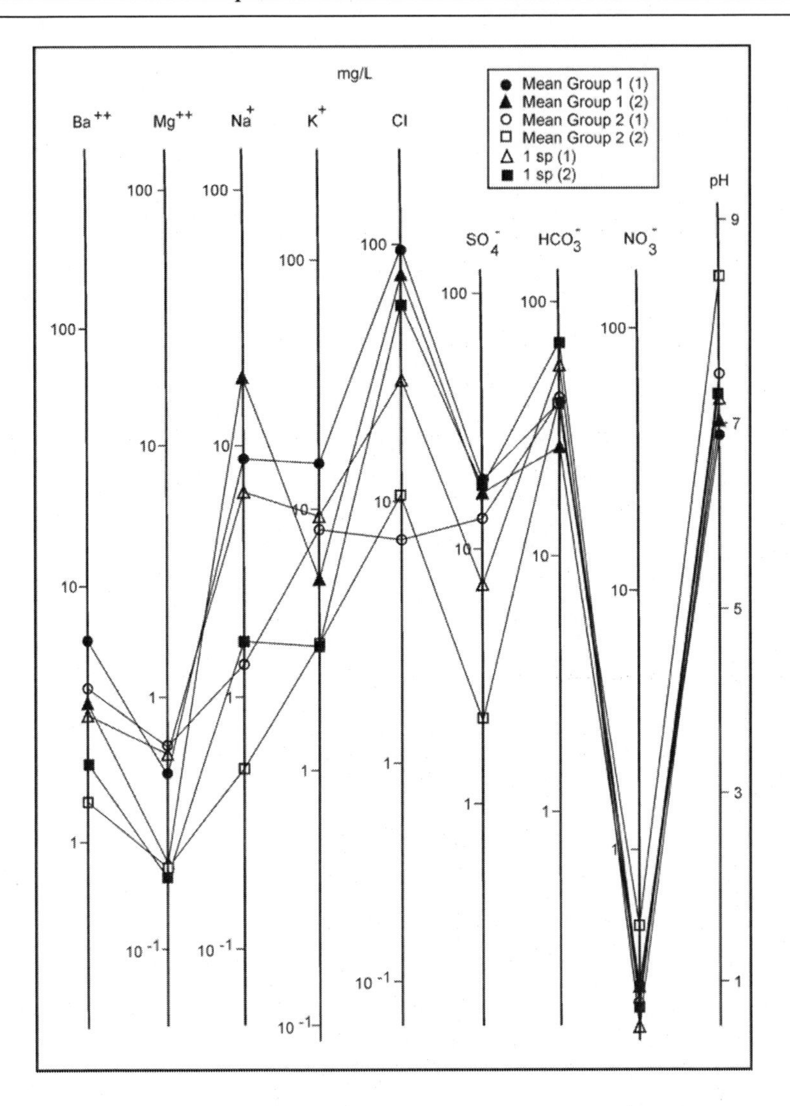

Figure 11. The chemical data for waters of Barreiro area plotted in a modified Schoeller-Berkaloff diagram.

The average values obtained for cations and anions in waters of Groups 1 and 2 plus sampling point 1sp were plotted in a modified Schoeller-Berkaloff diagram (Figure 11). Similar hydrochemical characteristics are verified within each group when the values are compared between the two field campaigns. In general, there is a concentration decrease for most cations and anions during the second field trip (dry period). The comparison of both groups indicates that most of the mean values obtained in Group 1 are higher than those in Group 2 for the two field campaigns. Group 1 waters tend to be Na^+-Cl^- dominated in the two campaigns, whereas Group 2 waters are K^+-HCO_3^- dominated. Figure 11 also indicates that point 1sp exhibits an intermediate composition, thus, characterizing a mixture of waters from the two drainages originating the Sal stream.

Figure 12 displays the results obtained in the chemical analysis of waters at point 5sp during the two field campaigns. It is evident that the high chloride content is possibly associated with anthropogenic pollution. This sampling point appears to exert a strong influence in the point 4sp, mainly in terms of the Cl content, since the waters of Grande Hotel lake (point 5sp) discharge into Cascatinha stream (point 4 sp).

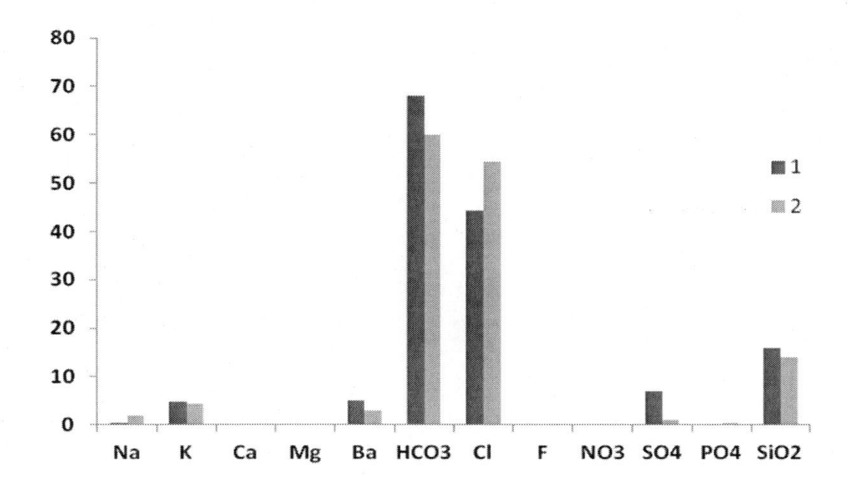

Figure 12. The chemical data (in mg/L) for waters of Barreiro area from monitoring point 5sp. 1 - February 2000; 2 - September 2000.

Figures 13 and 14 show the results obtained in the pH and Eh readings during the two field campaigns. The data are inserted within a transition zone, tending to exhibit a reducing character. The comparison of waters in Groups 1 and 2 indicates a different behavior among the drainages in the first field trip (Figure 13), contrarily to the second field trip (Figure 14). The Group 1 waters in the first sampling campaign tend to fit a better straight line than the samples of the second field trip. The Group 2 waters tend to fit straight lines whose slopes modify from the first to the second field campaign. These features reinforce the different characteristics and behavior of the drainages, and are also related to seasonal effects.

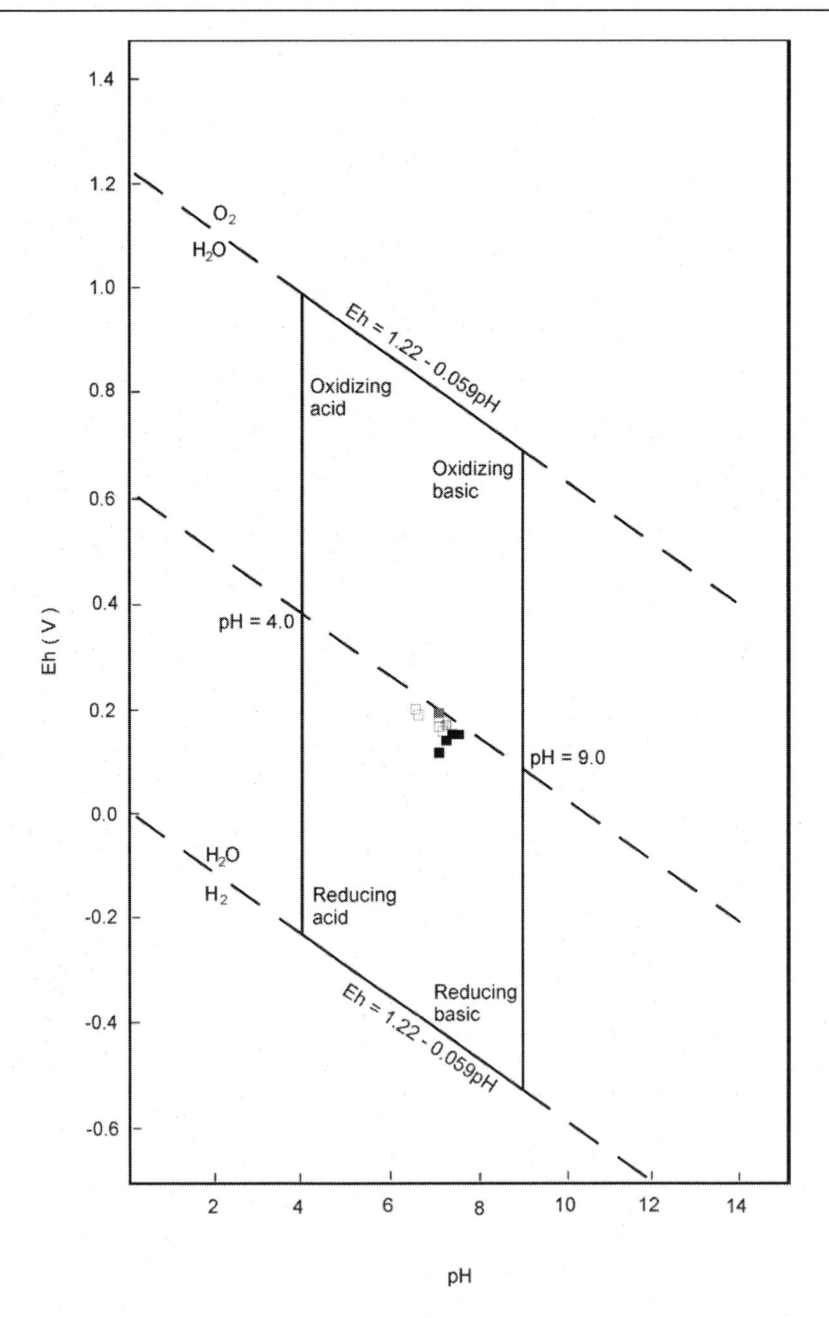

Figure 13. The pH and Eh data for Group 1, Group 2, 1sp and 5sp waters of Barreiro area plotted in an Eh-pH diagram. Sampling 1 - February 2000.

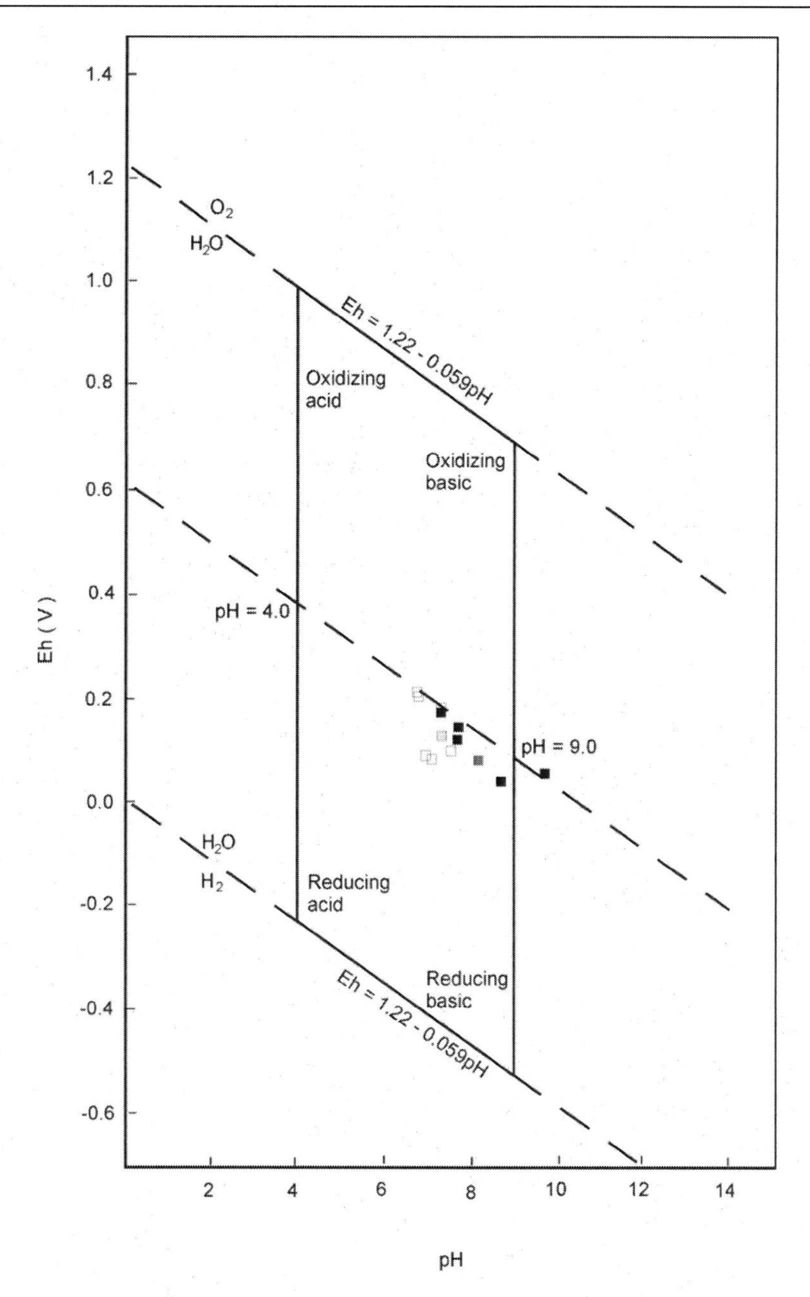

Figure 14. The pH and Eh data for Group 1, Group 2, 1sp and 5sp waters of Barreiro area plotted in an Eh-pH diagram. Sampling 2 - September 2000.

The hydrochemical data have been subjected to statistical analysis, where correlation coefficients were evaluated considering two criteria, i.e., geographical (drainage locations) and climatological (rainy and dry periods). The eastern waters (Group 1) exhibit the following significant correlations during the rainy period: conductivity *vs.* Ba (r = 0.91), conductivity *vs.* Na (r = 0.88), sulfate *vs.* Ba (r = 0.91), sulfate *vs.* Na (r = 0.96), sulfate *vs.* SiO_2 (r = 0.94), and Eh *vs.* pH (r = -0.98).

The Eh is inversely related to pH, as also shown in the Eh-pH diagram (Figure 13). The Group 1 waters exhibit the following significant correlations during the dry period: conductivity *vs.* Na (r = 0.97), conductivity *vs.* Cl (r = 0.99), sulfate *vs.* Na (r = 0.89), sulfate *vs.* Cl (r = 0.90), and Cl *vs.* Na (r = 0.97). Non-significant correlation was found between Eh and pH in waters of this period, as also indicated in the Eh-pH diagram (Figure 14). Thus, the correlations vary in the eastern waters between the rainy and dry periods.

The western waters (Group 2) only exhibit significant correlation between the Eh and pH during the rainy period (r = 0.87), as also shown in the Eh-pH diagram (Figure 13). However, the Eh was inversely related to pH in the dry period (r = -0.92), as confirmed by the Eh-pH diagram (Figure 14).

CONCLUSION

The hydrochemical features in each drainage suffer influence of the various mineral deposits occurring at Barreiro area, as well of the mining activities taking place there, mainly due to CBMM that possessed one effluents lagoon of the niobium extraction, which was situated upstream of Mata stream. The Group 1 drainage is located along the main mineral deposits of Barreiro area. Baritina stream originates at the niobium and phosphate mines, and CBMM stream flows through the barite deposits, whereas Mata stream and tributaries are distributed along the lateritic covers, occurring a rare earth elements deposit at its left margin and suffering influence of the effluents dam linked to the niobium mining. The Group 2 drainage is situated in an area that dominates the lateritic soil. The majority of the Cascatinha stream flow occurs in this material, reaching “F”dam, whereas Borges stream reaches “E”dam. The installations of the “BCM”dam and phosphate mine dyke also can influence the composition and quality of the surface waters, retaining or supplying certain compounds.

ACKNOWLEDGMENTS

FAPESP (Foundation Supporting Research in São Paulo State) and CNPq (National Council for Scientific and Technologic Development) in Brazil are thanked for financial support of some studies described in this chapter.

REFERENCES

[1] W. G. Mook, *Environmental isotopes in the hydrological cycle*, UNESCO-IAEA, Paris-Vienna (2000).

[2] L. E. Aragón and M. Clusener-Godt, Issues of local and global use of water from the Amazon, UNESCO, Montevideo (2004).

[3] F. Brandini, *Mar dulce mar*, http://www. amazonia.org.br/opiniao /artigo_detail.cfm?id=166380.

[4] DNPM (Departamento Nacional de Produção Mineral), *Principais depósitos minerais do Brasil*, DNPM, Brasília (1987).

[5] F. K. Dannemann, *Dona Beja – Araxá – Minas Gerais*, http://www.recantodasletras.com.br/resenhas/1677087

[6] Wikipedia, Dona Beija, http://pt.wikipedia.org/wiki/Dona_Beija

[7] G. Traversa, C. B. Gomes, P. Brotzu, N. Buraglini, L. Morbidelli, M. S. Principato. S. Ronca and E. Ruberti, *An. Acad. Bras. Ci.* 73, 71 (2001).

[8] J. R. K. Braga and H. Born, *Características geológicas e mineralógicas da mineralização apatítica de Araxá*, Anais Congresso Brasileiro de Geologia, 35, Belém (1988).

[9] A. Issa Filho, P. R. A. S. Lima and O. M. Souza, *Aspectos da Geologia do Complexo Carbonático do Barreiro, Araxá, MG*, CBMM (Companhia Brasileira de Metalurgia e Mineração), São Paulo (1984).

[10] L. A. Takata, J. M. N. Assis, C. A. Ikeda Oba, M. S. Cassola and H. Kahn, *Caracterização tecnológica de tipos de minério de fosfato da jazida da Arafértil, Araxá (MG)*, Anais Encontro Nacional de Tratamento de Minérios e Hidrometalurgia, 14, Salvador (1990).

[11] J. H. Grossi Sad and N. Torres, *Geology and mineral resources of the Barreiro Complex, Araxá, Brazil*, Anais Simpósio Internacional de Carbonatitos, 1, Poços de Caldas (1976)..

[12] L. O. Castro and J. M. Souza, *Estudo de urânio e terras raras associadas ao nióbio de Araxá – MG*, IPR (Instituto de Pesquisa Radioativas), Belo Horizonte (1970).

[13] R. S. Lopes, *Águas Minerais do Brasil (Composição, valor e indicações terapêuticas)*, Ministério da Agricultura, Rio de Janiero (1956).

[14] D. A. C. Beato, H. S. Viana and E. G. Davis, *Avaliação e diagnóstico hidrogeológico dos aqüíferos de águas minerais do Barreiro do Araxá, MG – Brasil*, Proc. Joint World Congress on Groundwater, 1, Fortaleza (2000).

[15] Funtec (Fundação Centro Tecnológico de Minas Gerais), *Relatório diagnóstico do conflito ecológico incluindo obras e medidas recomendadas para atenuação do impacto ecológico da mineração*, ECOS – Geologia, Consultoria e Serviços Ltda., Araxá (1984).

[16] Hach, *Water Analysis Handbook*, Hach Co., Loveland (1992).

[17] L. H. Mancini, *O rádio no ambiente hídrico do Morro do Ferro, Poços de Caldas (MG)*, M.Sc. Disssertation, UNESP, Rio Claro (1997).

[18] E. M. Tonetto, O tório em águas subterrâneas de Águas da Prata (SP), M.Sc. Disssertation, UNESP, Rio Claro (1996).

In: Basins
Editor: Jianwen Yang

ISBN: 978-1-63117-510-7

Chapter 7

CLIMATE DOWNSCALING OVER THE DENSU BASIN AND NEXUS ON HYDROLOGY IN GHANA, USING REGCM4

Raymond Abudu Kasei[1,2*]***, Barnabas Amisigo***[2]
and Boateng Ampadu[1]
[1]University for Development Studies -Earth and Environmental Science Department (EES), Faculty of Applied Sciences (FAS),
[2]International Water Management Institute, Accra, Ghana

ABSTRACT

Climate Downscaling is a term adopted in climate science in recent years to describe a set of techniques that relate local to regional-scale climate variables in relation to the larger scale atmospheric forcing. Theoretically, the techniques are advancements of the known traditional techniques in synoptic climatology. Climate downscaling specifically addresses the detailed temporal and spatial information from Global Climate Models (GCMs) required by precise researches of today. The Regional Climate Model version 4 (RegCM4), with a horizontal resolution of 55 km, was used to downscale the ECHAM5 simulations forced with observed SSTs over southern Ghana. An ensemble of 10 ECHAM5 AGCM integrations forced with observed time-evolving SSTs

[*] Corresponding author: University for Development Studies (UDS), P. O. Box 24, Navrongo, Upper East Region, Ghana, Telephone: +233 547076073, (E-mail: rakasei@gmail.com).

was done from 1961 to 2000. For each of the ECHAM5 AGCM integrations, a nested integration with the REGCM was done for the period January–June 1961–2000. Six-hour wind, temperature, humidity, and surface pressure data from ECHAM45 AGCM outputs were linearly interpolated in time and in space onto REGCM grids as base fields. The results of the comparison for the Densu catchment station showed a good correlation between the observed REGCM-simulated monthly rainfalls with significant statistics. Although no coherent trends were found in the basin, interannual rainfall variability was more pronounced as revealed by the REGCM 4 simulations. The northern part of the basin is most vulnerable to these variations because it has a monomodal rainfall pattern compared to the south which has relatively higher rainfall amounts due to its bi-modal rainfall pattern. The SPI analysis conducted on projected precipitation based on REGCM using IPCC's A1B and B1 scenarios against the base period of 1961-2000 showed both scenarios agreeing to a general drying trend for the future.

Keywords: Downscaling, climate, RegCM4, SPI, GCM

INTRODUCTION

Of all the Earth's resources, none is more fundamental to life than water. The restless atmosphere is an active agent in the constant redistribution of water on the earth's surface, which becomes even more striking when we realize that only a minute fraction of one percent of the earth's water is contained in the atmosphere at any time. For regions within West Africa, water resources are regarded as the life-blood of the economies of almost all countries. Over 70% of the inhabitants of West Africa depend primarily on rainfed agriculture for their livelihood. Hydropower is the main source of electric power generation until recently, crucial for socio-economic development, and is strongly dependent on availability of rainfall. Changes in the amount and distribution of rainfall have significant impacts on water availability and thereby directly influence socio-economic activities in the region. The Densu watershed within Ghana of West Africa is a catchment characterized by distinctive interannual and inter-decadal variability in precipitation. The availability of water in the watershed is of major importance for agriculture, as well as for industrial use and domestic use, impounded by a dam near the Ghana's capital, Accra. Due to increasing population pressure and corresponding intensification of agriculture and expansion of the capital city of Accra, the competition for water resources between these different sectors has intensified. Sustainable water management in Ghana requires scientifically sound estimates of future water availability. In particular, the impacts of global warming on precipitation distribution and quantity are of major concern for policy makers. It is therefore essential (a) to know if, and to what extent, global warming has already established its footprints in the Densu basin of West Africa, (b) to estimate future spatial and temporal rainfall distribution, and (c) to estimate future surface and subsurface water availability.

Dynamical downscaling is one of the approaches used to investigate the precipitation variability at local scales. A high-resolution limited-area model is run for the region of interest, forced by the large-scale circulation prescribed from a lower-resolution atmospheric global circulation model (AGCM). The strategy underlying this technique is that AGCMs can provide the response of global circulation to large-scale forcing, and nested regional models can account for the effects of local, sub-AGCM grid-scale forcings (Giorgi and Mearns 1999). Regional models have been tested for climate downscaling over North America, South America, Africa, Asia, and Europe (Fennessy and Shukla 2000; Giorgi and Marinucci 1991; Hong et al. 1999; Kanamtisu and Juang 1994; Nobre et al. 2001; Roads 2000; Seth and Rojas 2003; Sun et al. 1999a, b; Takle et al. 1999; Ward and Sun 2002). The simulation of both finer spatial-scale details and overall monthly or seasonal mean precipitation were improved in the regional models. Examination of other atmospheric variables indicated that the regional model simulations were generally as good as or better than those from AGCMs alone. Finer spatial-scale features developed in regional models are attributed to four types of sources: (i) the surface forcing such as detailed topography, (ii) the nonlinearities presented in the atmospheric dynamical equations, (iii) hydrodynamic instabilities (Denis et al. 2002), and (iv) noises generated at the lateral boundaries. Studies that concern analysis of daily precipitation in climate models are relatively few. Mearns et al. (1990) examined several versions of the AGCMs and found errors of too high and too low daily variability of precipitation depending on the version investigated.

A study by Gershunov and Barnett (1998) indicated that the AGCM missed important aspects of the El Nino Southern Oscillation (ENSO) signal in seasonal statistics of daily precipitation, although it is capable of capturing the ENSO signal in seasonal averaged precipitation which is reported to have some influence on the local climate of regions including West Africa. The analysis of daily precipitation in AGCMs probably is of limited value, given the crude horizontal resolution (i.e., the model can neither resolve important topographic influences on precipitation, nor synoptic-scale precipitation processes) and the crude parameterizations of precipitation (Mearns et al. 1995).

However, the model parameterization schemes are steadily improving, and regional models have relatively fine horizontal resolution. This mitigates some of the limitations of AGCMs, and examination of daily precipitation may prove more fruitful. Giorgi and Marinucci (1996) investigated the sensitivity of simulated precipitation to model resolution and concluded that the effect of model resolution on daily precipitation statistics is evident. Several studies indicate that high resolution regional models appear to be capable of improving the statistics of high-frequency precipitation toward more realism on finer spatial and time scales (Chen et al. 1999; Hong and Leetmaa 1999; Mo et al. 2000; Sun et al. 1999a). Although dynamical downscaling provides enhanced details of climate simulations, there is a need for more research to examine the regional model ability in simulating observed local climate conditions and evaluate the statistical structures of climate signals at various spatial and temporal scales to answer the question of whether predictability of climate systems is improved with regional modeling. A logical approach is to use multiple AGCMs with multiple ensembles and force multiple regional models (Leung et al. 2003).

STUDY SITE

The Densu and Akosombo Basins, which cover an area of about 5,000 km^2, have a high population density of about 387 persons/km^2, five times that of the national average of 77 persons/km^2. It is recognized as one of the most urbanized basins in the country. Apart from supplying water from its Weija reservoir to the over 400,000 people living in the western parts of Accra, it is a major source of water supply to the urban settlements of Koforidua, Suhum, and Nsawam with a combined population of about 140,000. Excess flow from the Weija reservoir discharges into the Densu delta (Sakumono) lagoon and salt pans complex, which constitutes one of Ghana's internationally recognized protected areas (Ramsar sites). A relatively fair observational network (approximate 12 stations) provides a good dataset to validate a fine scale downscaled results over the region (Figure 1).

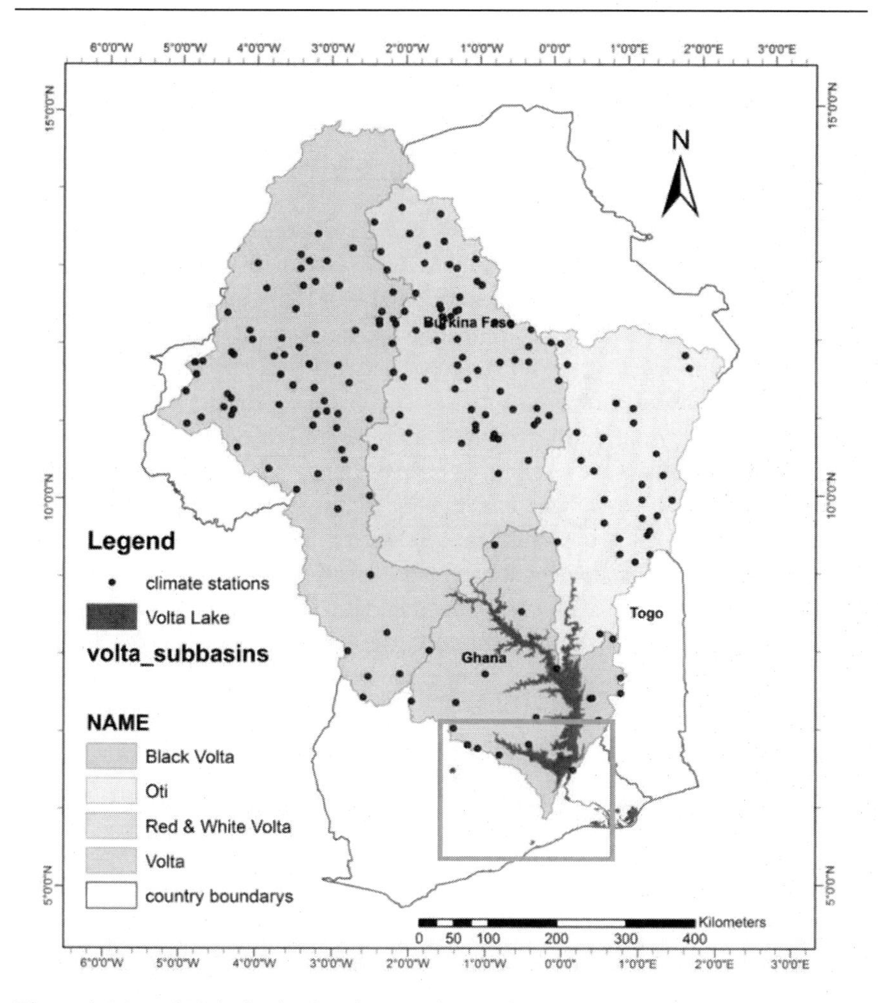

Figure 1. Map of Volta basin showing area interpolated using combined IDW for selected grids of 12 stations.

METHODOLOGY

Observed Trends

A statistical analysis of precipitation and temperature time series for Ghana was performed to identify current climatic trends within the whole country, of which the Densu basin was part. Available Ghana time series of annual and monthly maxima and corresponding monthly means were compiled from daily temperature and precipitation data. Additionally, trends in total annual precipitation were evaluated. Data gaps in temperature time series were filled by linear interpolation. Precipitation data gaps were not filled. The number of missing days and the number of days that could not be interpolated were counted. If the number of missing days exceeded a predetermined threshold, the corresponding month and year, respectively, were deleted from the time series. In the trend-analyses reported here, only time series containing at least 25 years of usable data are included. The Mann-Kendall test was used to determine the level of significance of assumed linear trends. Temperature time series showed predominantly positive trends. Most of the trends identified were statistically significant. Therefore, a conclusion that temperatures in Ghana are likely to be increasing can be verified. Precipitation time series on the other hand exhibited both negative and positive trends. Relatively few trends were significant, however. The significant trends were negative, with few exceptions. For annual precipitation (totals and monthly means), all significant trends were negative. However, only a relatively small number of trends among those examined were statistically significant, hence the conclusion that a clear trend towards decreasing precipitation is observed for Ghana using the Mann-Kendall test.

Experimental Design and Data

Several reports have shown that contemporary trends are inadequate scenario analyses, therefore, climate simulations were conducted to ascertain possible future climate characteristics. Dynamical downscaling through a regional climate model (RCM) was performed to deduce the regional impact of global climate change from the results of global climate models (GCM). Due to their coarse resolution, global climate simulations do not always provide sufficiently detailed potential future climate situations on a regional scale.

The regional model used in this study is the Regional Climate Model version 4 (RegCM4), which has evolved from the first version developed in the late 1980s (RegCM97; Dickinson et al. 1989; Giorgi 1990) to later versions in the early 1990s (RegCM2; Giorgi et al. 1993a,b), late 1990s (RegCM2.5; Giorgi and Mearns 1999), and 2000s (RegCM3; Pal et al. 2007). The RegCM97 is credited to be the first limited area model developed for long-term regional climate simulation, with which numerous regional model intercomparison projects have been applied by a large community for a wide range of regional climate studies, from basic process studies (Qian 2008; Qian et al. 2010) to advanced paleo-climate studies and future climate projections (Giorgi and Mearns 1999; Giorgi et al. 2006), as cited by Giorgi et al. 2012. The RegCM system is designed for use by a varied community composed of scientists in industrialized countries as well as developing nations (Pal et al. 2007). It is open source, user friendly and portable code that can be applied to any region of the world including the Densu catchment under study. RegCM is supported through a Regional Climate Research Network (RegCNET) which is a widespread network of scientists coordinated by the Earth System Physics section of the Abdus Salam International Centre for Theoretical Physics (ICTP; Giorgi et al. 2006). The evolved RegCM4 is the same as the basic dynamics as in RegCM3, which was essentially the same as that of the previous version RegCM2 (Giorgi et al. 1993a, b). RegCM4 is thus a hydrostatic and compressible.

The dynamical core was upgraded to the hydrostatic version. The REGCM97 was used to downscale ECHAM4.5 AGCM simulations. The one-way nesting of the REGCM97 into the ECHAM4.5 AGCM was done in a way that was different from conventional methods, which use global model results along the lateral boundary zone only. The perturbation nesting method used allows the global model outputs to be used over the entire regional domain DΦ3 (Figure 2), not just in the lateral boundary zone. The dependent variables in the regional domain are defined as a summation of perturbation and base. The base is a time-dependent prediction from the AGCM. All other variability that cannot be predicted by the AGCM but can be resolved and predicted by the REGCM in the regional domain is defined as a perturbation or adopted GLOWA Volta's MM5. Nesting is done in such a way that the perturbation is nonzero inside the domain but zero outside the domain.

The RegCM4 model was applied using a domain (DΦ3) having a horizontal resolution of 55 × 55 km^2 and 26 vertical layers extending up to 30 mbar at the model top, as used by the GLOWA Volta project that has carried out similar analysis within the region (Figure 2).

Feedback between soil moisture, temperature, vegetation, soil properties and atmosphere was taken into account by linking RegCM4 bi-directionally with the Oregon State University-Land Surface Model (OSU-LSM) (Chen & Dudhia, 2001). The OSU-LSM utilizes four soil layers down to 2 m depth to simulate energy and water fluxes at the atmosphere - land interface. The coupled OSU-LSM enables the simulation of feedback effects among soil, vegetation and atmosphere via calculation of soil- and vegetation-dependent sensible-, latent- and ground-heat fluxes. Elevation, land-use and soil data are taken from NCAR data archives and from data sets compiled by GLOWA-Volta project and other previous studies. Precipitation difference between REGCM simulations and the driving AGCM outputs is attributed not only to the different resolution forcing but also to the difference in the convection schemes used. It is necessary to verify if the REGCM can reproduce the large-scale information of precipitation in the driving AGCM.

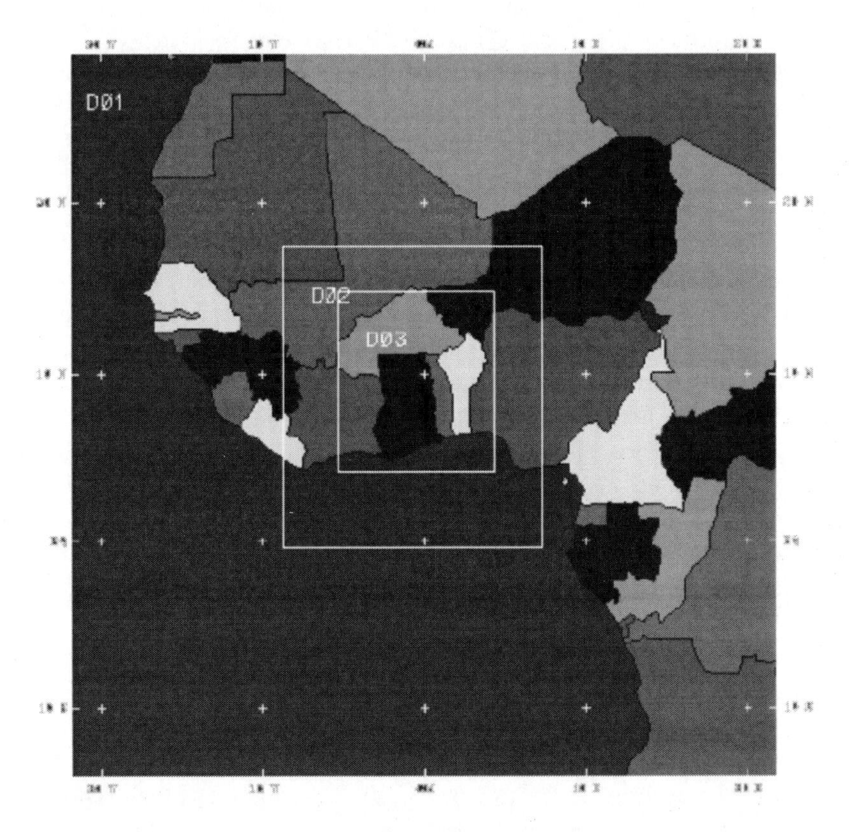

Figure 2. IPCC climate scenario of ECHAM5 was dynamically downscaled to the Densu and Akosombo catchments with RegCM4 using 55 × 55 km^2 resolution.

The choice of the regional model domain and resolution is very important when setting up the downscaling experiment (Giorgi and Mearns 1999). Several test simulations were performed to determine the optimum horizontal resolution and the size of the computational domain against the availability of computer resource. The simulation period was for the entire year of January–

December. Densu wet year of 1997 and dry year of 1983 were chosen for the test simulations. Compared with observations, the simulated seasonal precipitation bias over Densu is the smallest with model horizontal resolution of 55 km and the domain defined in Figure 1 (gridding of area of interest). Thus, this configuration is chosen for our downscaling study.

The test simulations show that the Atlantic ITCZ is very critical for the precipitation simulation in Densu. The entire tropical Atlantic Ocean has to be included in the domain. The displacement of the ITCZ occurs when only a portion of tropical Atlantic Ocean is included in the domain. With the resolution of 55 km, the main topographical features are resolved by the REGCM.

Our downscaling study was focused on the major rain season in Densu. An ensemble of 10 ECHAM5 AGCM integrations forced with observed time-evolving SSTs was done from the early 1960s to present. For each of the ECHAM5 AGCM integrations a nested integration with the REGCM was done for the period January–June 1970–2000. Six-hour wind, temperature, humidity, and surface pressure data from ECHAM45 AGCM outputs were linearly interpolated in time and in space onto REGCM grids as base fields. Soil wetness and soil temperature in REGCM were predicted after initialization.

Two experiments were carried out on dry and wet seasons over the catchments. The driest years derived from simulated rainfall were 1981, 1983, 1990, 1992 and 1998, whereas 1979, 1984, 1988, 1989, and 1991 were characterized by abundant monsoon rainfall. The results indicate that the model well simulates the rainfall intensity and distribution, and the circulation features during the extreme wet and dry seasons. For bias corrections, daily gauged data, CRU monthly and CPC daily data were used, in particular for model output statistics corrections on precipitation. The post processing variables of Precipitation (mm), Temperature (min, max, mean) (^{o}C), Relative humidity, Wind speed (m/s) and radiation (W/m^{2}) at daily resolution are analyzed.

RESULTS

Within the adopted REGCM, the spatial mean of the change of forest to agricultural land under the A2, B1 and A1B scenarios have been considered. For example, under the A1B scenario the estimate of the FAO (2006) assumes a decrease in forest coverage of about 30 % until 2050 for the entire Africa region. Associated albedo changes between 2000 and 2050 are in the order of 5-10 %. Forests transformation into grasslands and agricultural areas in the order of 10-15 % were incorporated. The B1 storyline and scenario family describes a convergence with the same global population, that peaks in mid-century and declines thereafter, as in the A1 storyline, but with a rapid change in economic structures toward a service and information economy, with reductions in material intensity and the introduction of clean and resource-efficient technologies. The emphasis is on global solutions to economic, social and environmental sustainability, including improved equity, but without additional climate initiatives. The A2 storyline and scenario family describes a very heterogeneous world. The underlying theme is self-reliance and preservation of local identities. Fertility patterns across regions converge very slowly, which results in continuously increasing population. Economic development is primarily regionally oriented, and per capita economic growth and technological change are more fragmented and slower than other storylines. The available REGCM outputs have three ensemble runs for the various scenarios and time slices. These are 901, 902 and 903 for the periods 1961-2000, the A1B scenarios with land-use changes are 911, 912 and 913 for 2001-2050 time periods, and the B1 scenarios are 921, 922 and 923 for 2001-2050 periods. Within the three ensemble runs of each scenario (Figure 3), no large differences were identified in annual totals between individual ensemble runs.

To assess the reliability REGCM future climate scenario for the evaluation of the impacts of climate change on water resources in the Densu Basin, the REGCM-simulated and another regional model developed by the GLOWA Volta project for the entire D3 domain, MM5-simulated mean rainfall for the time slice 1991-1997 obtained for the basin weather stations are compared to the mean observed rainfall for the same area and periods. This time slice was selected because it contained certified gauged meteorological data with minimal gaps. The results of the comparison for the Densu catchment station, for example, show a good correlation between the observed and MM5-simulated and REGCM-simulated monthly rainfall (Figure 4). Pearson correlation of gauged 1991-1997 and REGCM 1991-1997 is 0.823, and P-Value is 0.001; whereas for gauged 1991-1997 and MM5 1991-1997, the correlation is 0.957, and P-Value is < 0.0001. On average, MM5 and REGCM overestimate rainfall for this selected time slice of 1,203 mm and 1,322 mm per annum, respectively, against the measured 1,101 mm per annum. MM5 overestimates the rainfall from April through July and September and underestimates for August and October. REGCM, on the other hand, overestimates rainfall for February through April, July, October and November and underestimates for August. The strongest overestimation for MM5 is for the month of July; while for REGCM, it is March and April (Figure 4).

Although no coherent trends are found in the areas around the area of interest (AOI), interannual rainfall variability is more pronounced in the AOI as revealed by the REGCM protections (Figure 4). The northern part of the basin is most vulnerable to these variations because it has a monomodal rainfall pattern compared to the south which has relatively higher rainfall amounts due to its bi-modal rainfall pattern. The SPI analysis conducted on projected precipitation based on REGCM using IPCC's A1B and B1 scenarios against the base period of 1961-2000 (Figure 5) shows that both scenarios agree to a general drying trend for the future. With the exception of B1-scenario-based extreme dry year projections for 2016 (-2.5) and 2049 (2.1), the A1B scenario generally projects higher negative SPI values than B1, spanning the entire projection period and being more profound for the period 2019 to 2047.

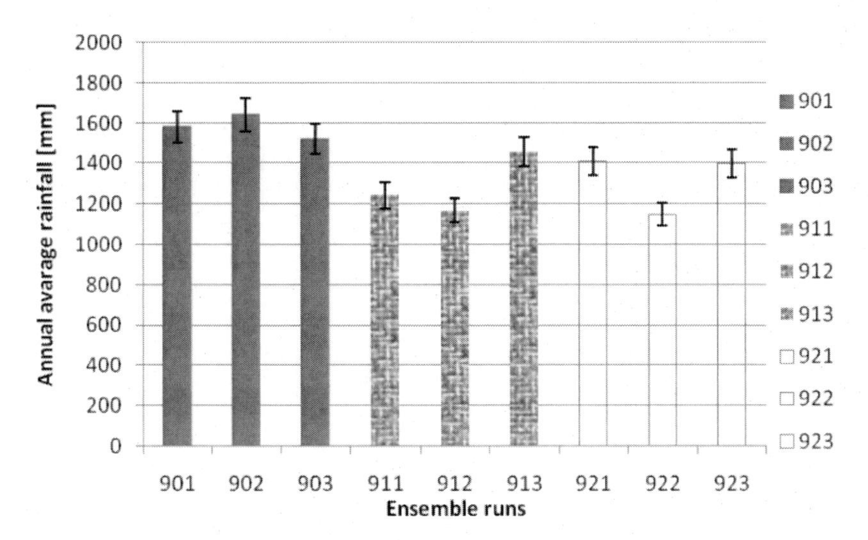

Figure 3. Annual rainfall average for three ensemble model runs of IPCC climate scenarios A2, A1B and B1 for present (1961-2000) and future (2001-2050), respectively, with standard deviations at Densu basin.

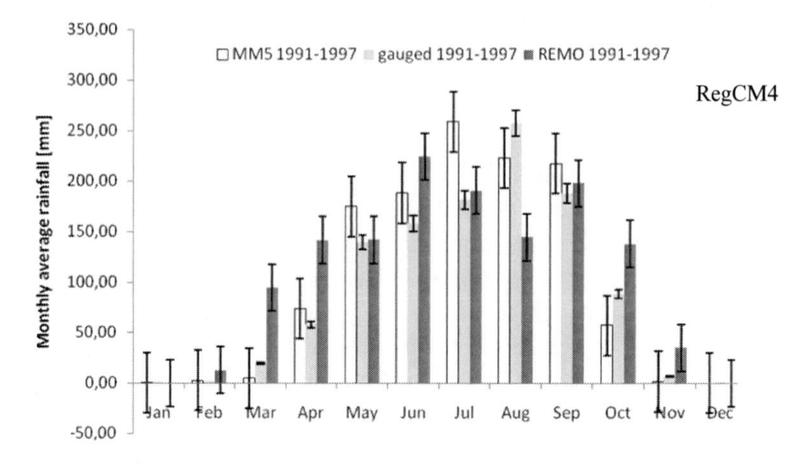

Figure 4. MM5-simulated (mean over 7 years) and RegCM-simulated (mean over 7 years) compared to observed (mean over 7 years of reliable) rainfall including standard deviation (error bars) rainfall at Densu basin.

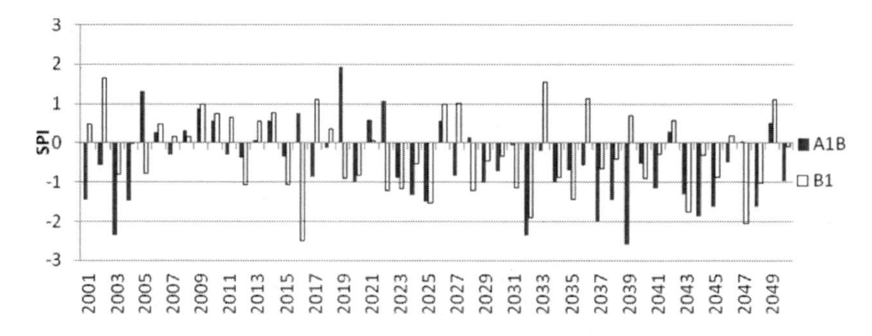

Figure 5. SPI characterization of REGCM simulations for A1B and B1 scenarios for 2001-2050 for Densu against base period of 1961-2000.

Both the scenarios have also projected few wet years for the future. For instance, both REGCM's A1B and B1 projections of 2008 and 2009 as very wet years have actually coincided with high rainfall and floods for the same years (official records). If these projections made in 2004 are anything to go by, then REGCM's projections of a blend of moderate and extreme dry years for the future should be given close attention in policy formulation. The SPI also indicates an increase in frequency of moderate to severe drought for the future (Table 1). MM5 does not have simulated outputs for 2008 and 2009 of the Densu Basin for validation.

Table 1. Comparison of climate occurrences of past (1961-2005 gauged) with future (2006-2050 REGCM's AIB-simulated) for the Densu Basin

	Past (gauged)	Future (REGCM-simulated)
	1961-2005 No. of occurrence	2006-2050 No. of occurrences
Severely-extremely wet	5	5
Moderate wet	5	5
Normal year	16	10
Moderate dry	11	6
Severely dry	4	12
Severely-extremely dry	3	6

CONCLUSION

Historical temperature time series show clear positive trends at high levels of statistical significance. Among precipitation time series, most significant trends are negative. Results of dynamic downscaling of ECHAM5 emissions scenario IPCC A1B and B1 show that in April, which is the usual transition from the dry to the rainy season, precipitation will decrease by up to 70% and the duration of the rainy season will narrow, which may have extensive implications for agriculture and city water supply. The predicted total annual precipitation decreases only slightly (5-10%) for the future. The predicted temperature increase in the rainy season is up to 2°C.

REFERENCES

Giorgi, F. (2006). Climate change hot-spots. *Geophys. Res. Let.*, *33*, doi:10.1029/2006GL025734

Mearns, L. O., Giorgi, F., Whetton, P., Pabon, D., Hulme, M. & Lal, M. (2004). *Guidelines for use of climate scenarios developed from regional climate model experiments*, Tech. rep., Data Distribution Centre of the IPCC

Dickinson, R. E., Errico, R. M., Giorgi, F. & Bates, G. T. (1989). A regional climate model for the western United States. *Clim Change*, *15*, 383–422

Dickinson, RE., Henderson-Sellers, A. & Kennedy, P. (1993). Biosphere–atmosphere transfer scheme (BATS) version 1eas coupled to the NCAR community climate model. Tech Rep, National Center for Atmospheric Research Tech Note NCAR.TN-387+STR, NCAR, Boulder, CO

Giorgi, F. (1990). Simulation of regional climate using a limited area model nested in a general circulation model. *J Clim*, *3*, 941–963

Giorgi, F., Marinucci, M. R. & Bates, G. (1993a). Development of a second generation regional climate model. (RegCM2). I. Boundary layer and radiative transfer processes. *Mon Weather Rev*, *121*, 2794–2813

Giorgi, F., Marinucci, M. R., Bates, G. & DeCanio, G. (1993b). Development of a second generation regional climate model (RegCM2). II. Convective processes and assimilation of lateral boundary conditions. *Mon Weather Rev*, *121*, 2814–2832

Pal, J. S., Giorgi, F., Bi, X., Elguindi, N & others. (2007). Regional climate modeling for the developing world: the ICTP Reg - CM3 and RegCNET. *Bull Am Meteorol Soc*, *88*, 1395–1409

Qian, J. H. (2008). Why precipitation is mostly concentrated over islands in the maritime continent. *J Atmos Sci*, *65*, 1428–1441

Qian, J. H., Robertson, A. W. & Moron, V. (2010). Interactions among ENSO, the monsoons, and diurnal cycle in rainfall variability over Java, Indonesia. *J Atmos Sci*, *67*, 3509–3524

Giorgi, F. (2006). Regional climate modeling: status and perspectives. *J Phys IV*, *139*, 101–118

Giorgi, F., Huang, Y., Nishizawa, K. & Fu, C. (1999). A seasonal cycle simulation over eastern Asia and its sensitivity to radiative transfer and surface processes. *J Geophys Res*, *104*, 6403–6423

INDEX

D

E

F

G

M

N

O

P

Q

R

S

T

U

V

W

Y

Z